Statistical Approaches for Hidden Variables in Ecology

SCIENCES

Statistics, Field Directors – Nikolaos Limnios, Kerrie Mengersen

Statistics and Ecology, Subject Head – Nathalie Peyrard

Statistical Approaches for Hidden Variables in Ecology

Coordinated by
Nathalie Peyrard
Olivier Gimenez

iSTE

WILEY

First published 2022 in Great Britain and the United States by ISTE Ltd and John Wiley & Sons, Inc.

ISTE Ltd
27-37 St George's Road
London SW19 4EU
UK

www.iste.co.uk

John Wiley & Sons, Inc.
111 River Street
Hoboken, NJ 07030
USA

www.wiley.com

Library of Congress Control Number: 2021949076

British Library Cataloguing-in-Publication Data
A CIP record for this book is available from the British Library
ISBN 978-1-78945-047-7

ERC code:
PE1 Mathematics
PE1_14 Statistics
LS8 Ecology, Evolution and Environmental Biology

MIX
Paper from responsible sources
FSC® C013604

Contents

Introduction

Nathalie PEYRARD[1], Stéphane ROBIN[2] and Olivier GIMENEZ[3]

[1] *University of Toulouse, INRAE, UR MIAT, Castanet-Tolosan, France*
[2] *Paris-Saclay University, AgroParisTech, INRAE, UMR MIA-Paris, France*
[3] *CEFE, University of Montpellier, CNRS, EPHE, IRD, Paul Valéry Montpellier 3 University, France*

I.1. Hidden variables in ecology

Ecology is the study of living organisms in interaction with their environment. These interactions occur at individual level (an animal, a plant), at the level of groups of individuals (a population, a species) or across several species (a community). Statistics provides us with tools to study these interactions, enabling us to collect, organize, present, analyze and draw conclusions from data collected on ecological systems. However, some components of these ecological systems may escape observation: these are known as hidden variables. This book is devoted to models incorporating hidden variables in ecology and to the statistical inference for these models.

The hidden variables studied throughout this book can be grouped into three classes corresponding to three types of questions that can be posed concerning an ecological system. We may consider the identification of groups of individuals or species, such as groups of individuals with the same behavior or similar genetic profiles, or groups of species that interact with the same species or with their environment in a similar way. Alternatively, we may wish to study variables which can only be observed in a "noisy" form, often called a "proxy". For example, the presence of certain species may be missed as a result of detection difficulties or errors (confusion with another species), or as a result of "noisy" data resulting from technology-related measurement errors. Finally, in the context of data analysis, we may wish to reduce the dimension of the information contained in data sets to a small number of explanatory variables. Note the shift from the notion of a variable which escapes observation, in the first cases, to a more generalized notion of hidden variables.

Statistical Models for Hidden Variables in Ecology,

All three of these problems can be translated into questions of inference concerning variables which, in statistical terms, are said to be latent. Inference poses statistical problems that require specific methods, described in detail here. The ecological interpretation of these variables will also be discussed at length. As we shall see, while the statistical treatment of these variables may be complex, their inclusion in models is essential in providing us with a better understanding of ecological systems.

I.2. Hidden variables in statistical modeling

The term "hidden variable", widely used in ecology, finds its translation in the more general notion of latent variables in statistical modeling. This notion encompasses several situations and goes beyond the idea of unobservable physical variables alone. In statistics, a latent variable is generally defined as a variable of interest, which is not observable and does not necessarily have a physical meaning, the value of which must be deduced from observations. More precisely, latent variables are characterized by the following two specificities: (i) in terms of number, they are comparable to the number of data items, unlike parameters that are fewer in number. Consider, for example, the case of a hidden Markov chain, where the number of observed variables and latent variables is equal to the number of observation time steps; (ii) if their value were known, then model parameter estimation would be easier. For example, consider the estimation of parameters of a mixture model where the groups of individuals are known.

In practice, if a latent variable has a physical reality but cannot be observed in the field (e.g. the precise trajectory of an animal, or the abundance of a seedbank), it is often referred to as a hidden variable (although both terms are often used interchangeably). In other cases, the latent variable naturally plays a role in the description of a given process or system, but has no physical existence. This is the case, for example, of latent variables corresponding to a classification of observations into different groups. We will refer to them as fictitious variables. Finally, latent variables may also play an instrumental role in describing a source of variability in observations that cannot be explained by known covariates, or in establishing a concise description of a dependency structure. They may result from a dimension reduction operation applied to a group of explanatory variables in the context of regression, as we see in the case of the principal components of a principal component analysis.

The notion of latent variables is connected to that of hierarchical models: if they are not parameters, the elements in the higher levels of the model are latent variables. It is important to note that the notion of latent variables may be extended to cover the case of determinist quantities (represented by a constant in a model). For example, this holds true in cases where the latent variable is the trajectory of an ordinary differential equation (ODE) for which only noisy observations are available.

I.3. Statistical methods

Some of the most common examples of statistical models featuring latent variables are described here.

Mixture models are used to define a small number of groups into which a set of observations may be sorted. In this case, the latent variables are discrete variables indicating which group each observation belongs to. Stochastic block models (SBMs) or latent block models (LBMs, or bipartite SBM) are specific forms of mixture models used in cases where the observations take the form of a network. Hidden Markov models (HMMs) are often used to analyze data collected over a period of time (such as the trajectory of an animal, observed over a series of dates) and take account of a subjacent process (such as the activity of the tracked animal: sleep, movement, hunting, etc.), which affects observations (the animal's position or trajectory). In this case, the latent variables are discrete and represent the activity of the animal at every instant. In other models, the hidden process itself may be continuous. Mixed (generalized) linear models are one of the key tools used in ecology to describe the effects of a set of conditions (environmental or otherwise) on a population or community. These models include random effects which are, in essence, latent variables, used to account for higher than expected dispersions or dependency relationships between variables. In most cases, these latent variables are continuous and essentially instrumental in nature. Joint species distribution models (JSDMs) are a multidimensional version of generalized linear models, used to describe the composition of a community as a function of both environmental variables and of the interactions between constituent species. Many JSDMs use a multidimensionsal (e.g. Gaussian) latent variable, the dependency structure of which is used to describe inter-species interactions.

In ecology, models are often used to describe the effect of experimental conditions or environmental variables on the response or behavior of one or more species. Explanatory variables of this kind are often known as covariates. These effects are typically accounted for using a regression term, as in the case of generalized linear models. A regression term of this type may also be used in latent variable models, in which case the distribution of the response variable in question is considered to depend on both the observed covariates and non-observable latent variables.

Many methods have been proposed for estimating the parameters of a model featuring latent variables. From a frequentist perspective, the oldest and most widespread means of computing the maximum likelihood estimator is the expectation–maximization (EM) algorithm, which draws on the fact that the parameters for many of these models would be easy to estimate if the latent variables could be observed. The EM algorithm alternates between two steps: in step E, all of the quantities involving latent variables are calculated in order to update the estimation of parameters in the second step, M. Step E focuses on determining the conditional distribution of latent variables given the observed data. This calculation may be immediate (as in the case of mixture models and certain mixed models) or possible but costly (as in the case of HMMs); alternatively, it may be impossible for combinatorial or formal reasons.

The estimation problem is even more striking in the context of Bayesian inference, as a conditional distribution must be established not only for the latent variables, but also for parameters. Once again, except in very specific circumstances, precise determination of this joint conditional law (latent variables and parameters) is usually impossible.

The inference methods used in models with a non-calculable conditional law fall into two broad categories: sampling methods and approximation methods. Sampling methods use a sample of data relating to the non-calculable law to obtain precise estimations of all relevant quantities. This category includes the Monte Carlo, the Markov chain Monte Carlo (MCMC) and the sequential Monte Carlo (SMC) methods. These algorithms are inherently random, and are notably used in Bayesian inference. Methods in the second category are used to determine an approximation of the conditional law of the latent variables (and, in the Bayesian case, of parameters) based on observations. This category includes variational methods and their extensions. These approaches vary in terms of the measure of proximity between the approximated law and the actual conditional law, and in terms of the distribution family used when searching for the approximation.

I.4. Approach and structure of our work

This book provides an overview of recent work on statistical modeling and estimation in latent variable models for ecology. The different chapters illustrate the main principles described above. In some cases, they present statistical methods based on classical models and algorithms; in others, the focus is on developments from recent research in others. Each chapter addresses a specific ecological issue and a modeling approach to solving the problem, illustrated using one or more case studies.

Readers may also access the R code[1] in order to make use of the tools presented here, applied to their own data.

1 https://oliviergimenez.github.io/code_livre_variables_cachees/.

Most of the questions associated with the case studies presented here relate to the comprehension or description of systems. While the issue of forecasting and prediction is touched upon in some chapters, this subject lies outside the main scope of our work. The issue of missing data (i.e. values not observed in samples) is also not addressed either. Finally, note that this work is not an exhaustive summary of latent variable models, or of the inference methods and algorithms used with these models. Each chapter touches on the question of inference in relation to the selected model; readers wishing to explore the subject in greater depth may wish to consult the references provided.

This book is not designed to be read from front to back, but rather as a resource on which ecologists working with models or statisticians working in the field of ecology may draw. Chapters are arranged in order of ecological scale, from individuals up to ecosystems, providing an initial interpretive framework. Another approach would be to consider the nature of the hidden variable being modeled. One final approach would be to examine different statistical models: some models are used in several chapters, in connection with questions on different scales, and using different estimation methods.

Table I.1 gives a summary of the contents of the different chapters and is designed to help readers identify material which is of interest to them.

I.5. Directions for further perspectives

The examples described above, along with those presented in the following chapters, highlight the immense flexibility of latent variables models. These models, involving one or more latent layers, provide a rich framework for the description of complex dependency structures, and/or for the approximation of a mechanistic description of the phenomena involved.

However, it is important to note that the most sophisticated models are almost always the most complex in terms of inference. It would be wrong to assume that inference simply "happens", whatever the statistical approach (frequentist, Bayesian, etc.). At the time of writing, there is no fully generic approach suitable for use with all models, and this is unlikely to change in the near future. Even the best-established algorithms (EM, MCMC, etc.) require users to have a good understanding of the underlying principles in order to guide and control their behavior, and/or to adjust the algorithm as needed. This need for adjustment is clearly visible in the chapters of this book.

To conclude this introduction, we wish to highlight two areas for further research in ecology, drawing on statistical modeling of hidden variables, which are not covered in this book but which show promise: namely the integration (or combination) of data from multiple sources, and the use of participative scientific data.

Chapter	Scale	Model	Latent variable Nature	Domain
1	Individual	Hidden Markov chain	Hidden: position Fictitious: behavior	$\mathbb{R}$ $\{1,\ldots,K\}$
2	Individual	Hidden Markov chain	Fictitious: proximal signals Hidden: resource acquisition and allocation	$\{1,\ldots,K\}$ $\{1,\ldots,K\}$
3	Population	Hidden Markov chain	Hidden: population dynamics	$\{1,\ldots,K\}$
4	Population	Noisy ODE	Hidden: population size	$\mathbb{N}^+$
5	Metapopulation	Spatialized hidden Markov chain	Hidden: class of dormant state	$\{1,\ldots,K\}$
6	Community	Mixture (SBM and bipartite SBM)	Fictitious: groups of species with the same interaction structure	$\{1,\ldots,K\}$
7	Community	Regression (JSDM for presence–absence)	Fictitious: correlations between species	$\mathbb{R}$
8	Community	Regression (JSDM on counts)	Fictitious: correlations between species	$\mathbb{R}$
9	Community	Regression (JSDM on counts)	Instrumental: components summarizing covariates	$\mathbb{R}$
10	Socio-ecosystem	Regression (SEM)	Hidden: components of the system which are not directly measurable and are in interaction	$\mathbb{R}$

Table I.1. *Chapters and contents*

Several works have recently been published on the integration of data from multiple sources in the field of ecology (Miller *et al.* 2019; Isaac *et al.* 2020). The aim of the authors is to systematically improve the precision of estimated data, potentially decreasing sample size, and to enable the estimation of parameters that cannot be approximated by any other means. Data integration generally involves a hierarchical modeling approach in which the hidden variable is present in all of the sources used in its estimation.

Data from participative scientific activity has also received increasing attention in the literature in recent years (Dickinson *et al.* 2012; McKinley *et al.* 2017). This is due to the increasing availability of the data, and to the fact that information can now be collected across an increasingly broad spatial and temporal scale. Participative data sources are a fascinating subject of study in statistical ecology, raising a number of new challenges in terms of spatial bias in sampling, or variations in participant expertise. Once again, a clear distinction between the ecological processes embodied by the hidden variables and the associated observation methods is essential in order to develop a full response to the ecological question.

I.6. References

Dickinson, J.L., Shirk, J., Bonter, D., Bonney, R., Crain, R.L., Martin, J., Phillips, T., Purcell, K. (2012). The current state of citizen science as a tool for ecological research and public engagement. *Frontiers in Ecology and the Environment*, 10(6), 291–297.

Isaac, N.J., Jarzyna, M.A., Keil, P., Dambly, L.I., Boersch-Supan, P.H., Browning, E., Freeman, S.N., Golding, N., Guillera-Arroita, G., Henrys, P.A., Jarvis, S., Lahoz-Monfort, J., Pagel, J., Pescott, O.L., Schmucki, R., Simmonds, E.G., O'Hara, R.B. (2020). Data integration for large-scale models of species distributions. *Trends in Ecology & Evolution*, 35(1), 56–67.

McKinley, D.C., Miller-Rushing, A.J., Ballard, H.L., Bonney, R., Brown, H., Cook-Patton, S.C., Evans, D.M., French, R.A., Parrish, J.K., Phillips, T.B., Ryan, S.F., Shanley, L.A., Shirk, J.L., Stepenuck, K.F., Weltzin, J.F., Wiggins, A., Boyle, O.D., Briggs, R.D., Chapin, S.F., Hewitt, D.A., Preuss, P.W., Soukup, M.A. (2017). Citizen science can improve conservation science, natural resource management, and environmental protection. *Biological Conservation*, 208, 15–28.

Miller, D.A.W., Pacifici, K., Sanderlin, J.S., Reich, B.J. (2019). The recent past and promising future for data integration methods to estimate species' distributions. *Methods in Ecology and Evolution*, 10(1), 22–37.

1

Trajectory Reconstruction and Behavior Identification Using Geolocation Data

Marie-Pierre ETIENNE[1] and Pierre GLOAGUEN[2]

[1] *Institut Agro, Agrocampus Ouest, CNRS, IRMAR – UMR 6625, Rennes, France*

[2] *Paris-Saclay University, AgroParisTech, INRAE, UMR MIA-Paris, France*

1.1. Introduction

The study of movement in ecology has taken off in recent years, driven by questions relating to the determinisms of individual movement. Interest in the ecology of movement has been largely fueled by the emergence and development of GPS technologies over the last 20 years, helped along by the creation of numerous databases made up of individual trajectories. These observations, on fine spatial and temporal levels, can be used to study the behavior of individuals in relation to their living environment. A variety of trajectory models have been developed and applied with the aim of reconstructing these behaviors and understanding the underlying determinisms. In this chapter, we shall present two latent variable models, widely used in movement ecology for trajectory analysis. Each model corresponds to a specific objective: the reconstruction of real trajectories with the removal of any geolocation errors, and the identification of different behaviors in the course of movement.

1.1.1. *Reconstructing a real trajectory from imperfect observations*

Trajectory data are frequently marred by errors for a variety of reasons (satellite accessibility issues, geolocation errors, etc.). This results in noisy observations of the

Statistical Models for Hidden Variables in Ecology,
coordinated by Nathalie PEYRARD and Olivier GIMENEZ. © ISTE Ltd 2022.

real position of the animal, which is itself unknown. The *hidden variable is, therefore, the real position* and the observed variable is the noisy version. In Figure 1.1, we can see that some recorded positions of a Cape dolphin, tracked using the Argos system, are actually on land – a situation which is evidently improbable. This observation almost certainly corresponds to noisy data concerning the actual position of the tracked individual.

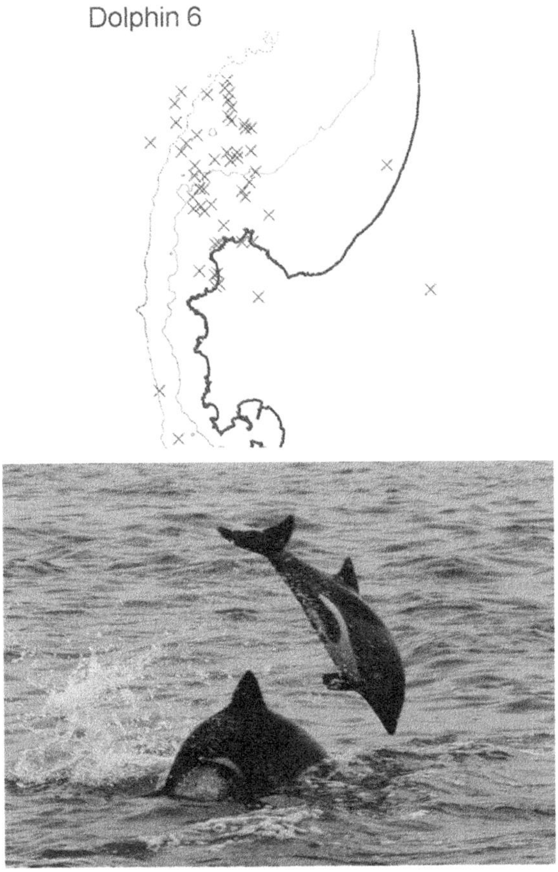

Figure 1.1. *The map at the top shows the tracking data for a male Cape dolphin (Cephalorhynchus heavisidii) in St. Helena Bay, South Africa. The coastline is shown in black, and we see that some recorded positions are actually on land. These positions are obtained using an Argos system. Figure taken from Elwen* et al. *(2006). Photo of a Cape dolphin by Jutta Luft, distributed under the GNU Free Documentation License. For a color version of this figure, see www.iste.co.uk/peyrard/ecology.zip*

Observation errors are generally small (a few meters) in cases where positions are obtained using a GPS system on open ground and with good satellite coverage. Far larger errors may occur using other technologies, such as the Argos system (into the tens of kilometers). A hierarchical model for reconstructing real trajectories from observed trajectories is presented in section 1.2.1.

1.1.2. *Identifying different behaviors in movement*

Individuals rarely move in a homogeneous manner, and different movement patterns are often observed. In Nathan *et al.* (2008), the authors propose a formalization of the mechanisms responsible for individual movement. Among the different aspects mentioned, the internal state of the individual and the environment in which it exists are identified as important mechanisms of movement. It seems reasonable to believe that the internal state of an individual affects its behavior, resulting in a change of movement regime.

Any study of individual movement must permit the identification of different states or activities. In this case, *the hidden variable is the activity of the individual*, while the observed variable is its position, or various metrics derived from this position, as we shall see later. Section 1.2.2 presents a reconstruction of behavior based on movement observations, using a specific latent variable model known as a hidden Markov model.

1.2. Hierarchical models of movement

1.2.1. *Trajectory reconstruction model*

1.2.1.1. *Overview*

In cases where there are errors in observed positions, data can be *smoothed* in order to recreate the real trajectory. To smooth errors, all collected data points are combined with a movement model in order to "straighten out" outlying observations and thus correct positioning errors.

Different ways of taking account of observation errors in movement models have been discussed at length in the literature (Freitas *et al.* 2008; Johnson *et al.* 2008; Patterson *et al.* 2010). For initial, simple trajectory reconstructions, however, a linear Gaussian hierarchical model can be used as a first data exploration. This approach draws on the notion that the observed position is a noisy version of the real position, and that the noise around this position is Gaussian. In formal terms, take n noisy observations, $y_{0:n} = (y_0, \dots, y_n)$, of an animal's position. Generally speaking (and throughout this chapter), we presume that each observed position is a vector of $\mathbb{R}^2$.

These observations are presumed to be realizations of random variables $Y_{0:n}$, the distribution of which depends on the real position of the animal. Moreover, the real position of an animal at a given instant (unknown) is dependent on its real position for the previous instant (also unknown). In formal terms, these positions themselves can be seen as a sequence of non-independent random variables, noted $Z_{0:n} = (Z_0, \dots Z_n)$, with values in $\mathbb{R}^2$.

We consider that all of these random variables obey the following hierarchical model:

$$\begin{cases} Z_0 \sim \mathcal{N}(\mu_0, \Sigma_0) & \\ Z_t = AZ_{t-1} + \mu + E_t^m, \quad E_t^m \overset{i.i.d}{\sim} \mathbb{N}(0, \Sigma^m), & 1 \le t \le n \\ Y_t = BZ_t + \nu + E_t^o, \quad E_t^o \overset{i.i.d}{\sim} \mathbb{N}(0, \Sigma^o) & 0 \le t \le n \end{cases} \quad [1.1]$$

From top to bottom, these three equations define:

– *The initial distribution*: the *a priori* initial position of the individual. In this case, we have a normal distribution (in dimension 2) about an initial position μ_0, with a variance–covariance matrix Σ_0.

– *The transition distribution* (or dynamic model): in this case, a model of the individual's movement. We consider that the current position is given by a random Gaussian variable, centered about an affine transformation of the previous position, with a variance–covariance matrix Σ^m. The affine transformation is obtained from two parameters: a matrix A (of size 2×2) and a vector μ of dimension 2. The most common approach is to consider that $\mu = \mathbf{0}$ and to take A as the identity matrix. The resulting model is a random walk.

– *The emission distribution* (or observation model): the observation is taken to be a random Gaussian variable centered about an affine transformation of the current position, with variance–covariance matrix Σ^o. The affine transformation is given by two parameters: a matrix B (of size 2×2) and a vector ν of dimension 2. The most common approach is to consider that $\nu = \mathbf{0}$ and to take B as the identity matrix. The observation is thus presumed to be centered about the real position.

1.2.1.2. *Inference*

Using the model defined by [1.1], inference is used for two purposes:

– *Estimation of positions*: in this case, inference is used to determine the distribution of actual positions based on observations, that is, for $0 \le t \le n$, the distribution of the random variable $Z_t | Y_{0:n}$. This distribution is known as the *smoothing distribution*.

– *Estimation of parameters*: to estimate the unknown parameters in the model (which, in the majority of cases, correspond to the two variance–covariance matrices, Σ^m and Σ^o).

With known parameters and for any $0 \leq t \leq n$, the distribution of $Z_t|Y_{0:n}$ is Gaussian. The mean and the variance–covariance matrix of this distribution can be calculated explicitly. This step is carried out using Kalman smoothing, which will not be described in detail here; interested readers may wish to consult Tusell (2011). It is important to note that the explicit nature of this solution is exceptional in the context of latent variable models, and is a result of the Gaussian linear formulation of model [1.1].

In practice, the parameter $\theta = \{\mu, A, \nu, B, \Sigma^m, \Sigma^o\}$ is unknown. In a frequentist context, the natural aim is to identify the parameter that maximizes the likelihood associated with observations $Y_{0:n}$:

$$L(Y_{0:n}|\theta) = \int \dots \int p(Y_{0:n}, x_{0:n}) \mathrm{d}x_0 \dots \mathrm{d}x_n,$$

where p is a generic notation for probability density. In this case, the expression of likelihood implies the calculation of an integral in very high dimensions, as it must be integrated across all hidden states. However, given a known sequence of real positions $X_{0:n}$, we would have an explicit expression of the *full log-likelihood*:

$$\begin{aligned} \ell\left(Y_{0:n}, X_{0:n}|\theta\right) &= \log p(X_0|\theta) + \log p(Y_0|X_0, \theta) \\ &+ \sum_{t=1}^{n} \left(\log p\left(X_t|X_{t-1}, \theta\right) + \log p\left(Y_t|X_t, \theta\right)\right). \end{aligned} \quad [1.2]$$

As all of the densities in this model are Gaussian, maximization of the log-likelihood would be simple. The *expectation–maximization* (EM) algorithm uses this full likelihood to maximize likelihood. Based on an initial parameter value $\theta^{(0)}$, the algorithm produces a series of estimations $\left\{\theta^{(\ell)}\right\}_{\ell \geq 0}$ as follows:

– Step **E** calculates:

$$Q(\theta|\theta^{(\ell)}) = \mathbb{E}\ell\left(Y_{0:n}, X_{0:n}|\theta\right)|Y_{0:n} = y_{0:n}, \theta^{(\ell)}. \quad [1.3]$$

– Step **M** takes:

$$\theta^{(\ell+1)} = \mathrm{argmax}_\theta Q(\theta|\theta^{(\ell)}).$$

The series $\{\theta^{(\ell)}\}_{\ell\geq 0}$ converges to a local maximum likelihood (Dempster *et al.* 1977). Equation [1.3] consists of calculating the expectation of [1.2] with respect to the distribution of missing data given the existing observations, that is, the distribution of $X_{0:n}|\{Y_{0:n} = y_{0:n})$, using a "real" parameter $\theta^{(\ell)}$. The smoothing distribution for the parameter $\theta^{(\ell)}$ must, therefore, be calculated as part of this step; this is done using Kalman smoothing. A solution is then obtained explicitly in step M thanks to the Gaussian linear nature of the problem.

1.2.1.3. *Filtering and smoothing a trajectory*

As we have seen, the reconstruction of a trajectory is reliant on the determination of a smoothing distribution, that is, for all $0 \leq t \leq n$, the distribution of $Z_t|Y_{0:n}$. Note that the inference of the real position at a time t takes account of *all observations*. As this distribution is Gaussian in the context of the model [1.1], this corresponds to calculating $\mathbb{E}[Z_t|Y_{0:n}]$ and $\mathbb{V}[Z_t|Y_{0:n}]$ using Kalman recursions.

The name Kalman is more often encountered in the context of Kalman filtering, rather than Kalman smoothing. In these contexts, the *Kalman filter* is used to determine the *filter distribution*, that is, the distribution of $Z_t|Y_{0:t}$. It is, thus, the distribution of the position at time t on the basis of the *observations* up to time t.

Intuitively, smoothing gives a better estimation than filtering, as the future can be taken into account when estimating a position at time t. Using filtering, the tth position is corrected using positions observed up until time t, while in the case of smoothing, all of the available information is taken into account. Figure 1.2, taken from Lopez *et al.* (2015), illustrates the advantages of smoothing. A precise, frequent recording of the movement of an elephant seal, obtained using GPS (the reference curve), is shown alongside a reconstruction of the same real trajectory obtained using Argos data, filtering and smoothing.

1.2.2. *Activity reconstruction model*

1.2.2.1. *Overview*

As we indicated earlier, an individual alternates between different activities, and these are reflected in different modes of movement. For example, an individual who is looking for food will move slowly, with frequent changes of direction as potential food sources are detected. An individual traveling back to the colony, on the other hand, will travel relatively quickly and in a relatively straight line.

Subjacent (hidden) activities may be reconstructed by analyzing a trajectory, using a model that connects activities and movement. In this case, the observations $y_{0:n} = (y_0, \dots, y_n)$ are measures of a metric, which is presumed to be affected by an animal's activity (typically, this metric represents speed; other examples are discussed in the following section). Taking $0 \leq t \leq n$, z_t is used to represent the unobserved activity of an individual at an instant t. This activity is encoded as an integer between 1 and J, where J is a known integer, representing the number of expected activities. Observations and hidden activities are considered as realizations of random variables. Let $\mathbf{Z} := (Z_0, \ldots, Z_n)$ be the series of hidden states (subjacent activities) and $\mathbf{Y} := (Y_0, \ldots, Y_n)$ the series of movement measurements.

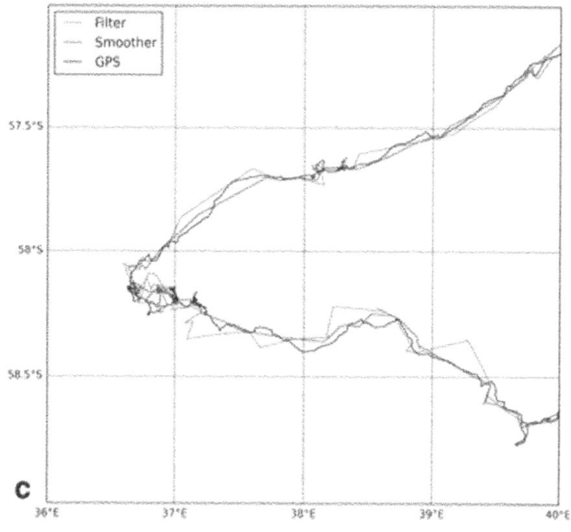

Figure 1.2. *Figure extracted from Figure 4 in Lopez* et al. *(2015). The black line shows a precise recording of the movements of an elephant seal. The green line was obtained by filtering positions recorded using the Argos system, and the purple line shows a smoothed version of the same data. The trajectory reconstructed using smoothing corresponds more closely to the reference data than the version obtained by filtering. For a color version of this figure, see www.iste.co.uk/peyrard/ecology.zip*

Using a classic activity reconstruction approach, the sequence $\mathbf{Z}$ is modeled by a Markov chain, that is, the series of random variables Z_t verifies the Markov property; in other terms, for any series of integers $z_{0:t}$ with values in $\{1, \ldots, J\}^{i+1}$:

$$\mathbb{P}\left(Z_t = z_t | Z_{0:(t-1)} = z_{0:(t-1)}\right) = \mathbb{P}\left(Z_t = z_t | Z_{t-1} = z_{t-1}\right).$$

Furthermore, if we consider that this probability of transition is independent of the instant t, the Markov chain is said to be *homogeneous*[1].

1 This hypothesis is helpful from a mathematical perspective, and results in efficient hidden state reconstruction algorithms. From a modeling perspective, its use is debatable, since it implies that the individual has no memory of its past trajectory.

The model draws on the idea that the distribution of Y_t is dependent on the activity. The modeler must, therefore, specify the distribution of $Y_t | \{Z_t = j\}$. This specification is generally carried out using a parametric distribution (typically a normal distribution). Activity identification is based on the ways in which the parameters of this distribution change (the mean and variance change as the activity changes).

The full model is formulated as follows:

$$\begin{array}{ll} \mathbb{P}(Z_0 = j) = \nu_0(j), & 1 \leq j \leq J \\ \mathbb{P}(Z_t = j | \{Z_{t-1} = i\}) = \Pi(i, j), & 1 \leq i, j \leq J \\ Y_t | \{Z_t = j\} \overset{i.i.d}{\sim} \mathcal{D}ist.(\theta_j) & 1 \leq j \leq j \end{array} \quad [1.4]$$

From top to bottom, these three equations define:

– *The initial distribution*: this the probability distribution for the first activity, and is thus a vector of probabilities $\nu_0 = (\nu_0(1), \ldots, \nu_0(J))$. In the common case where only one trajectory is observed, the initial distribution is taken to be known, or equal to a uniform distribution over $\{1, \ldots, J\}$.

– *The transition distribution*: in the case of a homogeneous Markov chain, the transition distribution is fully characterized by the matrix Π, of size $J \times J$, of which each line is a probability vector.

– *The emission distribution*: the observation is taken to be a random variable, the distribution of which depends, via these parameters, on the activity. The nature of the distribution depends on the nature of the observations. Note that observations are considered to be independent, conditionally to $\mathbf{Z}$.

This model is shown in the graphical form in Figure 1.3.

1.2.2.2. *Choice of observation metric*

This general framework offers many possibilities in terms of modeling. The subjacent activity may influence different aspects of the trajectory. For example, the trajectory of an individual looking for food will include multiple changes in direction. On the other hand, when an individual is traveling, in the context of migration, for example, its trajectory tends to be relatively straight with only minor changes in direction. In this example, changes in direction are strong activity markers.

Most of the metrics encountered in existing literature are based on the speed and direction of the animal in question.

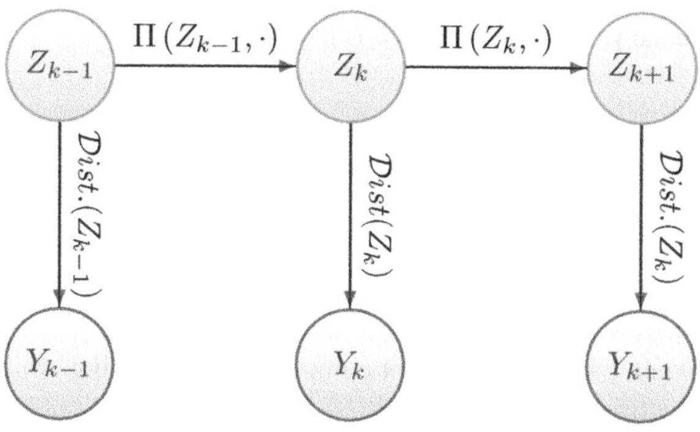

Figure 1.3. *Graphical model. For a color version of this figure, see www.iste.co.uk/peyrard/ecology.zip*

Starting from the positions $\{P_t\}_{t>0}$ (with values in $\mathbb{R}^2$ obtained at times $0, \Delta, 2\Delta, \dots$), the process of speeds $\{V_t\}_{t\geq 1}$ is defined by

$$V_t = \frac{P_t - P_{t-1}}{\Delta}.$$

From these speeds, we can define the direction $\{\psi_t\}_{t\geq 0}$ (with values in $[-\pi, \pi[$) as the angle between V_t and a reference vector (typically the vector (1, 0) pointing east). From these metrics, we deduce step length (or scalar speed) processes, denoted as$\{L_t\}_{t\geq 1}$, and turning angles, denoted as$\{\varphi_t\}_{t\geq 1}$ (with values in $]-\pi, \pi]$ using the convention $\varphi_1 = 0$) as follows:

$$L_t = ||V_t|| \quad [1.5]$$

$$\varphi_t = \begin{cases} \psi_t - \psi_{t-1} & \text{if } |\psi_t - \psi_{t-1}| \leq \pi \\ -2\pi + (\psi_t - \psi_{t-1}) & \text{if } \psi_t - \psi_{t-1} \geq \pi \\ 2\pi + (\psi_t - \psi_{t-1}) & \text{if } \psi_t - \psi_{t-1} \leq -\pi \end{cases} \quad [1.6]$$

The step length and turning angle metrics were the first to be used in behavior, or activity, analysis based on HMMs (Morales *et al.* 2004) and have been widely used (Patterson *et al.* 2008). In this way, we obtain the model illustrated in Figure 1.3, where Y_t is a bivariate vector of coordinates (L_t, φ_t) (often considered to be independent). One drawback to this method is the need to define an emission distribution, which is compatible with angles in order to model (φ_t). In practice, Von Mises (Jammalamadaka and Sengupta 2001) or Wrapped Cauchy distributions are the most widely used.

A different set of equivalent metrics may be used in order to avoid working with circular distributions, as proposed in Gurarie *et al.* (2009) and Gloaguen *et al.* (2015);

these are persistence velocity (V_t^p) and turning velocity (V_t^r):

$$V_t^p = L_t \cos(\varphi_t) \qquad [1.7]$$

$$V_t^r = L_t \sin(\varphi_t) \qquad [1.8]$$

An observation Y_t is thus a vector made up of these two components. As these components are signed, it is logical to model Y_t using a bivariate normal distributions. Where relevant, this model allows the introduction of a dependency relationship between the two movement components, something which is difficult to achieve when selecting a couple (L_t, φ_t).

Ecological expertise concerning the effects of different activities on movement can also contribute to the choice of an appropriate metric. In the case study presented at the end of this chapter, the two classic metrics were used to illustrate the difference between the two approaches, in terms of both results and practical implementation.

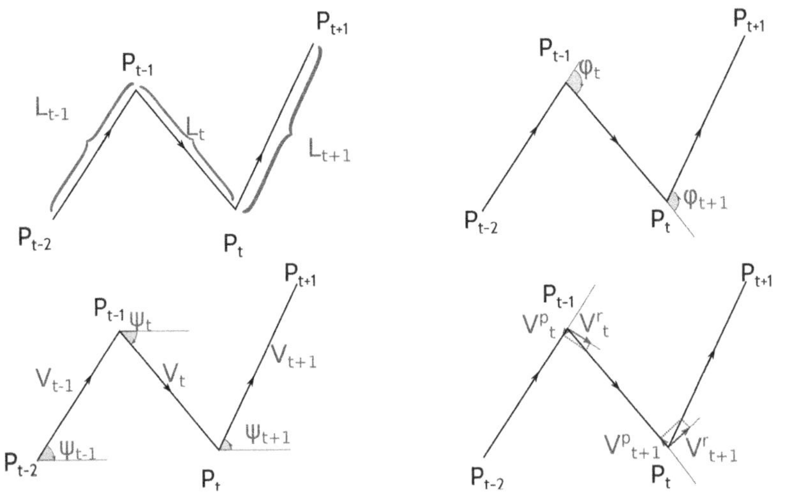

Figure 1.4. *Illustration of the quantities present in equations* [1.5]–[1.8]. P_t *denotes the successive positions occupied by the tracked individual. The series of speed vectors denoted as*(V_t) *and* (L_t) *denotes step length as defined by equation* [1.5]. *The series of directions is denoted as*(Ψ_t), *while* (ϕ_t) *is the series of turning angles as defined by equation* [1.6]. *For a color version of this figure, see www.iste.co.uk/peyrard/ecology.zip*

1.2.2.3. *Covariates inclusion*

A further question concerns the extent to which activity is influenced by covariates (distance from a point of interest, time of day, etc.). One way of including covariates is to model their impact on the transition between activities (Calenge *et al.* 2009; Morales *et al.* 2004; Michelot *et al.* 2016).

For example, in the model presented here, $\mathbb{P}(Z_t = j|Z_{t-1} = i)$ is independent of t and takes a value of $\Pi(i, j)$. Let us suppose that at each moment t, p covariates are measured and stored in a line vector $\mathbf{x}_t$. Transition probability can be linked to these variables according to a multiclass logistic regression approach:

$$\ln \frac{\mathbb{P}(Z_t = j|Z_{t-1} = i)}{\mathbb{P}(Z_t = i|Z_{t-1} = i)} = \mathbf{x}_t \beta(i, j), \quad \text{for all } j \neq i, \qquad [1.9]$$

$$\mathbb{P}(Z_t = i|Z_{t-1} = i) = 1 - \sum_{j=1, j \neq i}^{J} \mathbb{P}(Z_t = j|Z_{t-1} = i).$$

The first equation indicates that the probability of switching to a different activity j from a current activity i is connected to external conditions via a linear combination of covariates at time t. $\beta(i, j)$ is the column vector (of dimension p) of the coefficients corresponding to the influence of each covariate on this probability. The second equation is a constraint equation that ensures that the vector $(\mathbb{P}(Z_t = 1|Z_{t-1} = i)), \ldots, \mathbb{P}(Z_t = J|Z_{t-1} = i))$ is a probability vector.

It is thus possible to take account of notions such as the fact that an individual will spend a longer period of time actively foraging in a location that is rich in food sources, while in a less favorable environment, it will rapidly switch to a traveling state in order to move to a better location. The inclusion of covariates in this model makes it possible to identify environmental variables, which favor particular states.

1.2.2.4. *Example: a three-state HMM with Gaussian emission*

Let us illustrate model [1.4] using a toy example. Consider an individual with a total of three possible behaviors. Thus, let $J = 3$ be the number of activities for this individual. Each of these activities is characterized by a different movement pattern: for example, a direct trajectory at high velocity, a more sinuous pattern at a lower speed, and a third, different pattern. As we have seen, these differences may be characterized using different metrics. In this example, we have chosen to model persistent velocity and turning velocity, using equations [1.7] and [1.8].

Thus, in model [1.4], ν_0 is a probability vector of size 3, Π is a 3×3 matrix such that the sum of the elements in each line is equal to 1, and, for $1 \leq j \leq 3$, distribution (θ_j) is a distribution $\mathbb{N}(\mu_j, \Sigma_j)$, where:

– μ_j is a vector of dimension 2 (the mean of V^p and V^r for activity j);

– Σ_j is a variance–covariance matrix (of size 2×2).

1.2.2.5. *Inference*

Using the model defined by [1.4], inference is used to fulfill two purposes:

– *Estimation of activity*: to determine the distribution of real activities given the observations, that is, for $0 \leq t \leq n$, the distribution of the random variable $Z_t|Y_{0:n}$. For each time t, the estimated smoothing distribution gives the probability of being involved in each of the j activities.

– *Estimation of parameters*: the distribution ν_0 and the transition matrix Π characterize the dynamic of activities, and the set of parameters $\{\theta_j\} 1 \leq j \leq J$ indicates the way in which the activity influences the distribution of observations.

In the case of unknown parameters, these two steps are carried out conjointly. Taking a frequentist approach, the EM algorithm may be used, as in section 1.2.1. Once again, it is easy to write an equation, analogous to [1.2], giving the full likelihood.

For this model, step E once again consists of calculating the quantity given by [1.4]. Again, the difficulty lies in calculating the smoothing distribution. Nevertheless, the discrete character of the hidden dynamic means that explicit calculation is possible. This is carried out iteratively using the *forward–backward* algorithm. The equations used in this simple and efficient algorithm can be found in Rabiner (1989).

Step M, in which the parameters are updated, is dependent on the nature of the emission distribution. In the Gaussian case, this step has an explicit expression; conversely, this is not the case when using a distribution such as von Mises.

Bayesian estimation may be applied by using Monte Carlo Markov Chain (MCMC) algorithms, which are found in programs such as Stan (Carpenter *et al.* 2017), Winbugs (Lunn *et al.* 2000) or NIMBLE (de Valpine *et al.* 2017).

1.2.2.6. *Reconstruction of hidden states*

The reconstruction of hidden activities allows us to identify homogeneous phases in behaviors, and is often of considerable interest from an ecological perspective. This hidden Markov model may thus be seen as an unsupervised segmentation/classification model for movement.

One possibility is to reconstruct the most likely hidden activity for each time increment in turn, taking $\hat{Z}_{0:n} = (\hat{Z}_0, \dots, \hat{Z}_n)$ such that

$$\hat{Z}_t = \text{argmax}_j \mathbb{P}\left(Z_t = j|Y_{0:n}\right). \qquad [1.10]$$

This method is known as the *maximum a posteriori* (MAP) method. One possible problem with the estimator [1.10] is that it provides no guarantee that the sequence

$\hat{Z}_{0:n}$ will be coherent with the transition matrix Π; it may give a result of $\hat{Z}_t = 1$ and $\hat{Z}_{t+1} = 2$ for an estimation $\hat{\Pi}(1,2) = 0$, for example.

Using a Bayesian approach, the sampling algorithms used to estimate parameters permit the use of a *joint* smoothing distribution, that is, samples of $Z_{0:n}|Y_{0:n}$ can be obtained. Each sample produced is thus a possible sequence of activities corresponding to given observations.

Sampling across this distribution can also be carried out in conjunction with a frequentist approach, but the combinatorial level is high and the computational effort involved rapidly becomes prohibitive as n increases. Hidden activities are most commonly reconstructed using the most probable sequence of hidden states, that is, which maximizes the overall a posteriori distribution, or, more formally,

$$\hat{Z}_{0:n} = \underset{i_0, i_1, \ldots, i_n}{\arg\max}\, \mathbb{P}\left(Z_0 = i_0, Z_1 = i_1, \ldots Z_n = i_n | Y_{0:n}\right)$$

This sequence can be calculated in an efficient manner using the Viterbi algorithm, and is the version which is generally returned by libraries offering frequentist estimation. Note that the m most probable sequences can be obtained using the generalized Viterbi algorithm (Guédon 2007).

1.2.2.7. *Choosing the number of activities*

There are two very different approaches to choosing a number of behaviors or activities. The first is based on biological criteria, and a certain number of different behaviors may be identified. In the example of the red-footed booby, described later, a distinction is made between periods of rest, slow flight (corresponding to foraging) and rapid, direct flight, corresponding to trajectories between two points of interest.

Nevertheless, in the case of a new species or study environment, it can be hard to establish an initial idea of the number of hidden states; in this case, an approach based on statistical, rather than biological, criteria may be preferred. In statistics, this is known as a model choice problem, with "model" corresponding to a number of components.

One well-known model choice criterion is the Akaike information criteria (AIC) (Akaike 1973), which can be used to ensure that the number of parameters fits the data as well as possible. The aim is not simply to identify a parsimonious model, which fits the data; states need to be as different as possible, meaning that the problem is also one of classification. A new state should only be added if it is sufficiently distinct from other states. In this case, the integrated complete likelihood (ICL) criterion may be used (Biernacki *et al.* 2000; Bacci *et al.* 2014).

Given a set of estimated parameters for the model and a sequence of most probable states $\widehat{Z}$, reconstructed using the Viterbi algorithm, for example, this criterion is defined thus:

$$ICL = -2 \log \mathbb{P}_{\widehat{\Theta}} \left(Y_{0:n}, \widehat{Z}_{0:n} \right) + d \times \log n,$$

where $\hat{\Theta} = \left(\hat{\nu_0}, \hat{\Pi}, \left\{ \hat{\theta}_j \right\}_{1 \leq j \leq J} \right)$ and d is the total number of free parameters in the model in question. Using this definition, the model which minimizes the ICL will be selected.

1.3. Case study: masked booby, *Sula dactylatra* (originals)

The data used in this section were collected by Sophie Bertrand (IRD), Guilherme Tavares (UFRGS), Christophe Barbraud and Karine Delord (CNRS). The authors wish to thank the IRD Tabasco JEAI (Jeune Equipe Associée Internationale) for permission to use these data.

Figure 1.5. *Masked Booby (Sula dactylatra) Photo: Sophie Bertrand. For a color version of this figure, see www.iste.co.uk/peyrard/ecology.zip*

1.3.1. *Data*

This case study concerns the behavior of the masked booby (*Sula dactylatra*).

The data used here consist of three trajectories of three masked boobies around the Ilha do Meion, an island in the Fernando de Noronha archipelago (Brazil) in the Atlantic Ocean (Figure 1.6). Positions were recorded by GPS, with data collected every 10 s.

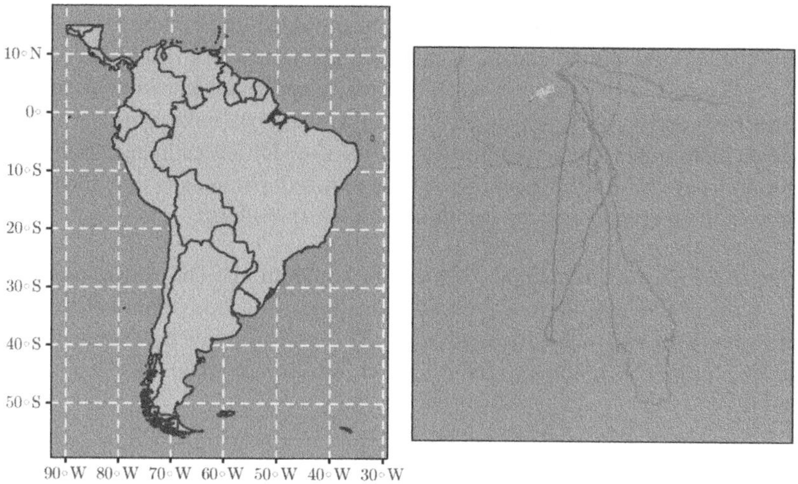

Figure 1.6. *Area of study (shown in red on the map) and three trajectories obtained by tracking three different red-footed boobies. Data were acquired in time increments of 10 s. For a color version of this figure, see www.iste.co.uk/peyrard/ecology.zip*

1.3.2. *Projection*

The recorded data for the boobies were provided in the form of latitude and longitude measurements, that is, in terms of angles with respect to an origin point on the Earth's surface. The methods presented earlier use a notion of distance (such as step length). While it is possible to calculate distances traveled over the Earth's surface using latitude and longitude coordinates, this requires the use of specific formulas for movement on a sphere. Instead, data are often projected onto a plane, enabling the use of Euclidean distance. Due to the spherical nature of the globe, the actual projection used depends on the zone of interest[2]. In this case, projection is carried out using UTM coordinates for zone 25 – south. In `R`, the `sf` library may be used to facilitate geographical data processing (and, notably, projection).

1.3.3. *Data smoothing*

In this case, the frequency of data acquisition was high (one point every 10 s). While there were few errors in the data (obtained using GPS), the temporal proximity of observations may result in a somewhat erratic-looking trajectory. This erratic effect is even more pronounced in the movement metrics used to detect different activities.

2 One degree of longitude at the equator does not correspond to the same distance as 1 degree of longitude at the 45th parallel, for example.

To correct errors, let us take a Gaussian linear hidden Markov model, as described in section 1.2.1. Taking the equations in model [1.1], matrices A and B are taken as known and equal to the identity, while vectors μ and ν are known and equal to 0. Matrices Σ^m and Σ^o are presumed to be diagonal, but unknown. The unknown variables, represented the actual position, in this model are estimated using an EM algorithm from the `MARSS` package. The estimated parameters are then used to reconstruct the real trajectory by means of Kalman smoothing.

Figure 1.7 shows an example of trajectory smoothing. This smoothing process greatly reduces the irregularities present in the trajectory. It is important to note that we are now working with *processed* data. This transformation is shown here for illustrative purposes, although its relevance in this specific case is somewhat debatable.

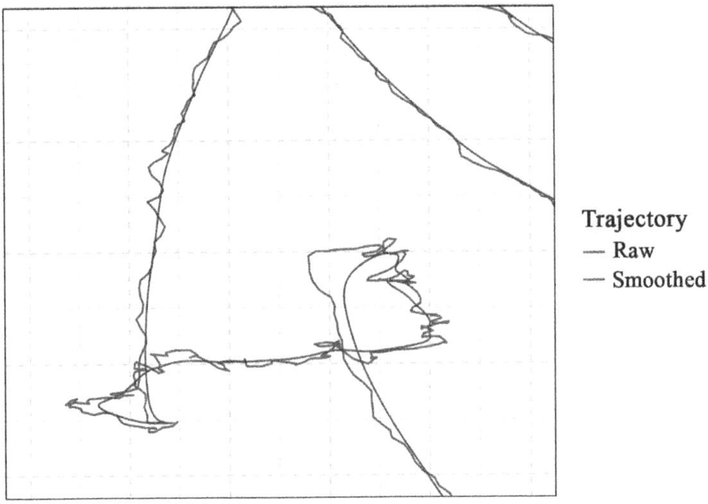

Figure 1.7. *Result of Kalman smoothing on part of the booby trajectories. Smoothing clearly removes many of the erratic aspects of the trajectory. For a color version of this figure, see www.iste.co.uk/peyrard/ecology.zip*

1.3.4. *Identification of different activities through movement*

We shall begin by using a three-state model. The choice of the number of states in this case will be discussed later.

1.3.4.1. *Definition of metrics*

Our aim is to identify different activities within a trajectory. In this case, we wish to distinguish between foraging behaviors (associated with rapid changes in direction) and, for example, direct movement toward a point of interest, which results in a straighter trajectory. In this case, turning angle and step length appear to be the most relevant metrics. Biological knowledge concerning the movement of these birds supports the use of these metrics to distinguish between different behaviors.

In this example, we have chosen to adjust two models, which differ in the way in which they treat step length and turning angles (and thus in the associated emission distributions). The two pairs of metrics considered here are as follows:

– Step length and turning angle: a classic choice, as presented by Morales *et al.* (2004): the emission distributions in this case are a gamma distribution for step length and a circular (von Mises) distribution for angles. This model will be labeled *length/angle* in our figures.

– Bivariate velocity change metric (Gurarie *et al.* 2009): the emission distributions in this case are two independent normal distributions. This model will be labeled *bivariate speed* in our figures.

1.3.4.2. *Defining the starting point of the algorithm*

These models do not include any covariates, and the initial distribution will not be estimated. Each model is made up of 18 parameters (12 emission distribution parameters and six transition matrix parameters). Iterative optimization applied to a space of this type (such as the EM algorithm) may be affected by the chosen starting point. In both cases, the choice of a suitable starting point for the algorithm is crucial. One relatively generic approach involves a classification of k-averages (for the selected metrics). This rapid classification can be used to identify plausible parameters for different regimes; nevertheless, it is still important to ensure that the result obtained from the algorithm has not been affected by the choice of starting point.

1.3.5. *Results*

1.3.5.1. *Characterization of hidden states*

In the two packages used here, the parameters of the HMM are estimated using maximum likelihood, and the sequence of most probable hidden states is retraced using a Viterbi algorithm.

The hidden states in this model characterize the distribution of speeds and turning angles. In terms of trajectories, this implies that a hidden state characterizes a segment between two positions (10 s apart, in this case). States on the trajectory are thus represented on these segments. In this unsupervised classification approach, the labels assigned to hidden states (interpreted as behaviors) are arbitrary, and each state must be characterized *a posteriori*. For ease of interpretation, we have chosen to categorize three states according to the average observed speed. State 1 corresponds to the activity with the lowest average speed, while state 3 corresponds to the fastest average speed.

Figure 1.8 shows the behaviors identified by inference along trajectories for the two models used here.

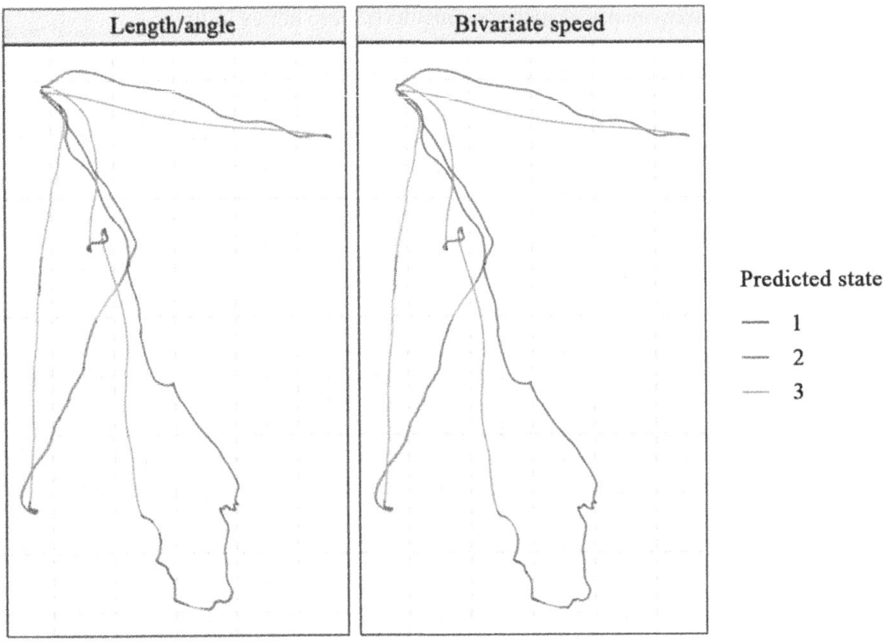

Figure 1.8. *Representation of states along trajectories estimated using two different models. States are classified in order of relative speed, from slowest (1) to fastest (3). For a color version of this figure, see www.iste.co.uk/peyrard/ecology.zip*

It is clear from the figure that the two models result in different nuances in the trajectories.

– Using the *length/angle* metric, three states are clearly visible, notably a "slow" state that characterizes the erratic phases of the trajectory. An intermediate state (2) characterizes trajectories of medium speed and a medium level of eraticness, while the third state reflects fast, direct movement.

– Using the *bivariate velocity*, metric, a similar third state is obtained, but there are significant differences in terms of the distinction between the first two states. In the first state, the bird appears to be "drifting", at slow speeds with little variation. The second state appears to characterize all movement during which the bird is not resting, with a high level of variation in terms of speed.

This initial interpretation can be extended by analyzing the distribution of step length and turning angles by state[3] (Figure 1.9).

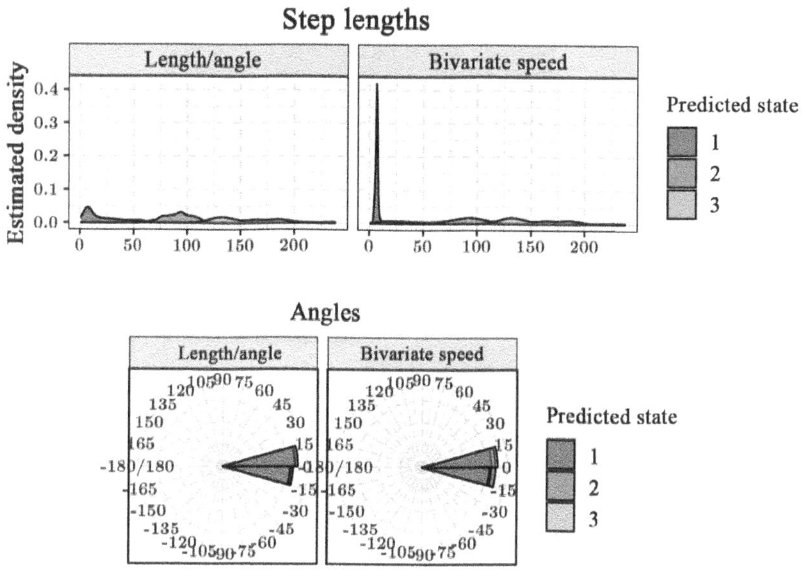

Figure 1.9. *Distribution of our chosen metrics for the states estimated using our two models. For a color version of this figure, see www.iste.co.uk/peyrard/ecology.zip*

It is immediately evident that the distinction between states is not based on the distribution of turning angles in either model. The influence of the step length distribution is much clearer.

We see that, for the *length/angle* metric, the slow state corresponds to movements of between 0 and 50 m; using the *bivariate velocity* metric, this same state corresponds to movement over very short distances (of the order of 5 m). Consequently, state 2 covers different speed intervals; using the first metric, this state corresponds to step lengths of 60–120 m, whereas for the second metric, this state corresponds to step

3 The "bivariate velocity" model can still be interpreted *a posteriori* using the classic step length/angle approach, as shown here. In our example, as the majority of turning angles are close to 0, the V_p component of the metric is closely correlated with the step length.

lengths of 0–100 m, including much shorter distances (of around 20 m). Conversely, step 3 (rapid movement) corresponds to similar movement patterns in both metrics.

This distinction can also be seen in the table of state contingencies by metric (Figure 1.10).

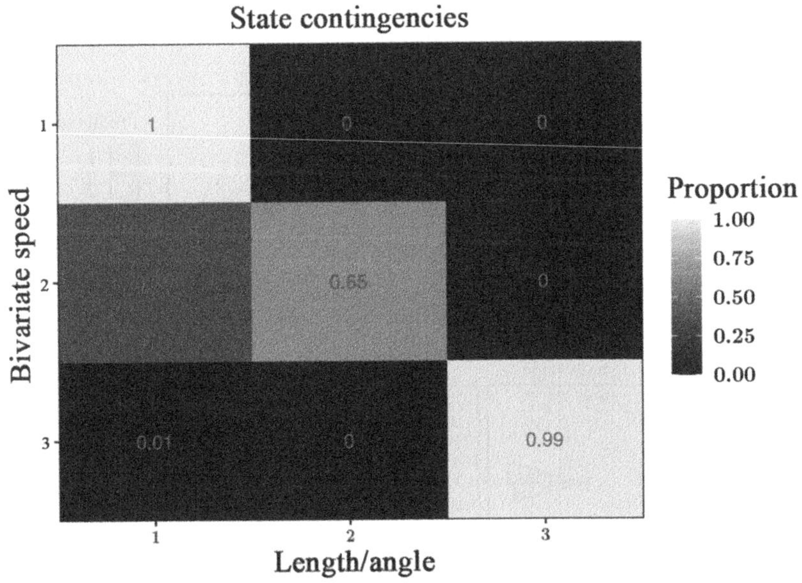

Figure 1.10. *Contingencies of estimated states for our two models. For a color version of this figure, see www.iste.co.uk/peyrard/ecology.zip*

Once again, we see that state 1 of the *bivariate velocity* metric falls within state 1 of the *Length/Angle* metric; similarly, the third states of the two metrics correspond. The metrics differ in the way in which state 2 is characterized: state 2 of the *Bivariate Velocity* metric corresponds to a combination of states 1 and 2 of the *Length/Angle* metric.

These differences are not surprising, given the differences between the underlying metrics. We have chosen to highlight this difference here for illustrative purposes, but it should be noted that, for a four-state model, the disparities are much smaller.

In this context, the characterization of states and the choice of the "best" model is based on interpretation, drawing on biological knowledge of the species in question. As is often the case, this unsupervised approach is most suitable for exploratory purposes, and should be interpreted in light of the ecological context.

1.3.5.2. *State uncertainty*

One advantage of the probabilistic version of supervised classification is the fact that the uncertainty of classification can be quantified. In this case, we can consider the evolution of the probability of being in state 1 or state 2 over time. Figure 1.11 illustrates this evolution for one of the three boobies.

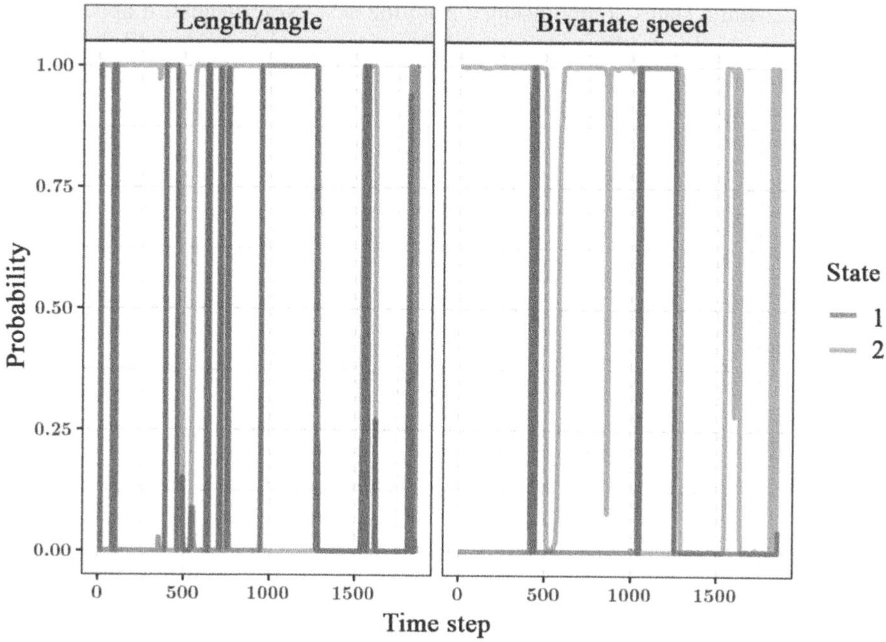

Figure 1.11. *Evolution of the probability of being in state 1 or state 2 over time, by model. For a color version of this figure, see www.iste.co.uk/peyrard/ecology.zip*

We see that the level of uncertainty in terms of state classification is very low, as the separation of distributions in each state is clear.

1.3.5.3. *Inclusion of the nest distance covariate*

Our model may be developed further by adding a covariate, in this case the bird's distance from the nest. We wish to determine whether changes in activity are correlated with distance from the nest. Using the formulation given in equation [1.9], we write:

$$\ln \frac{\mathbb{P}(Z_t = j | Z_{t-1} = i))}{\mathbb{P}(Z_t = i | Z_{t-1} = i))} = \beta_0(i,j) + \beta_1(i,j) d_t + \beta_2(i,j) d_t^2,$$

where d_t is the distance from the nest at time t. A quadratic component has been added to take account of the potentially variable character of the relation between the probability of transition and the covariate.

Figure 1.12 shows the evolution of the probability of state transitions as the distance from the nest changes. We see that, overall, there is a high level of persistence within states at any distance from the nest. Nevertheless, it appears that the closer the bird is to the nest (to the right on the x axis), the less likely it is to transition from activity 2 to activity 3. Note that the interest of the chosen covariate in this specific case is debatable; based on the AIC, a model without this covariate would be preferred.

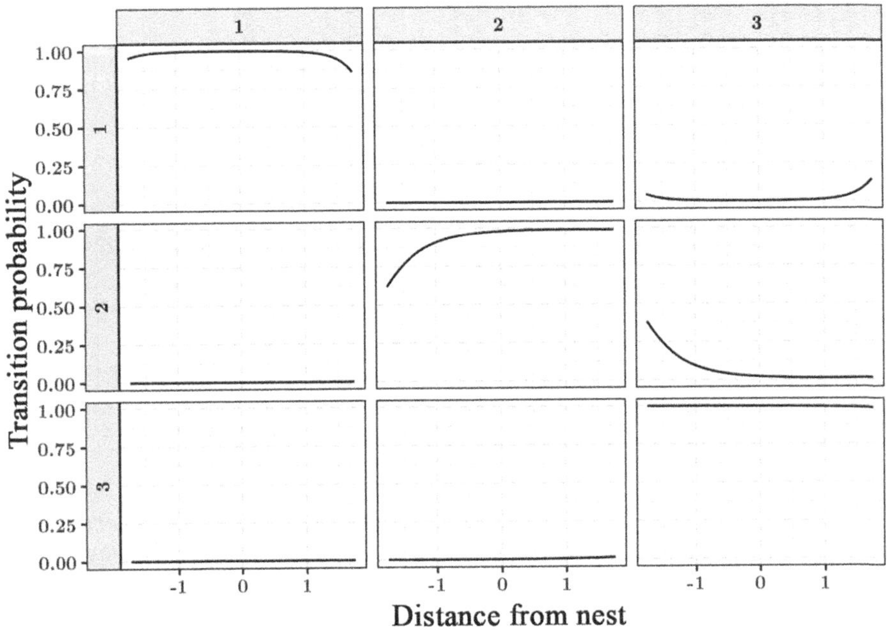

Figure 1.12. *Evolution of estimated transition probabilities as a function of distance from the nest. The figure should be read as a transition matrix. The graph in the second line, third column represents the evolution of the probability of a transition from state 2 to state 3 as a function of distance from the nest. As the distance variable has been centered and reduced, the origin represents the mean distance from the nest across all data points. For a color version of this figure, see www.iste.co.uk/peyrard/ecology.zip*

1.3.5.4. *Choosing a number of states*

The calculation of model selection criteria is valuable in helping to chose the number of states to use, as is the AIC. Table 1.1 shows AIC and ICL scores for different numbers of activities across our three trajectories.

J	2	3	4	5	6	7
AIC	29,044	24,213	18,773	16,624	**14,220**	19,480
ICL	29,195	24,210	18,887	16,720	**14,821**	21,003

Table 1.1. *Evolution of model selection criteria (AIC and ICL) as a function of the number of hidden states J. In both cases, the best scores are attained for a model with six hidden states*

From a purely statistical perspective, a 6-state model appears preferable here.

Figure 1.13 shows states along a trajectory (using the bivariate velocity model) alongside the speed characteristics of these states. We see that a classification into six activities broadly corresponds to the creation of subdivisions in the intermediate state. States previously characterized as belonging to activity 2 or 3 (Figure 1.9, top left) are divided into four different groups in the new model. In our view, the choice of an optimum number of states in this case should be guided by our capacity to interpret the model, rather than by purely statistical considerations.

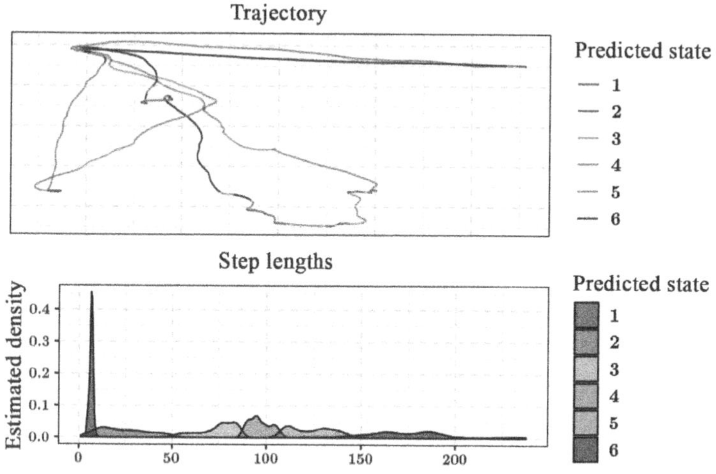

Figure 1.13. *Study zone (red dot on the map) and three trajectories of three different red-footed boobies. Measured over a time step of 10 s. For a color version of this figure, see www.iste.co.uk/peyrard/ecology.zip*

1.4. References

Akaike, H. (1973). Information theory as an extension of the maximum likelihood principle. In *Second International Symposium on Information Theory*, Petrov, B.N., Csaki, F. (eds). Akademiai Kiado, Budapest.

Bacci, S., Pandolfi, S., Pennoni, F. (2014). A comparison of some criteria for states selection in the latent Markov model for longitudinal data. *Advances in Data Analysis and Classification*, 8(2), 125–145.

Biernacki, C., Celeux, G., Govaert, G. (2000). Assessing a mixture model for clustering with the integrated completed likelihood. *IEEE Transactions on Pattern Analysis and Machine Intelligence*, 22(7), 719–725.

Calenge, C., Dray, S., Royer-Carenzi, M. (2009). The concept of animals' trajectories from a data analysis perspective. *Ecological Informatics*, 4(1), 34–41.

Carpenter, B., Gelman, A., Hoffman, M.D., Lee, D., Goodrich, B., Betancourt, M., Brubaker, M., Guo, J., Li, P., Riddell, A. (2017). Stan: A probabilistic programming language. *Journal of Statistical Software*, 76(1), 1–32.

Dempster, A., Laird, N., Rubin, D. (1977). Maximum likelihood from incomplete data via the EM algorithm. *Journal of the Royal Statistical Society*, 39(1), 1–38.

Elwen, S., Meÿer, M.A., Best, P.B., Kotze, P., Thornton, M., Swanson, S. (2006). Range and movements of female Heaviside's dolphins (Cephalorhynchus heavisidii), as determined by satellite-linked telemetry. *Journal of Mammalogy*, 87(5), 866–877.

Freitas, C., Lydersen, C., Fedak, M.A., Kovacs, K.M. (2008). A simple new algorithm to filter marine mammal Argos locations. *Marine Mammal Science*, 24(2), 315–325.

Gloaguen, P., Mahévas, S., Rivot, E., Woillez, M., Guitton, J., Vermard, Y., Etienne, M.-P. (2015). An autoregressive model to describe fishing vessel movement and activity. *Environmetrics*, 26(1), 17–28.

Guédon, Y. (2007). Exploring the state sequence space for hidden Markov and semi-Markov chains. *Computational Statistics & Data Analysis*, 51(5), 2379–2409.

Gurarie, E., Andrews, R.D., Laidre, K.L. (2009). A novel method for identifying behavioural changes in animal movement data. *Ecology Letters*, 12(5), 395–408.

Jammalamadaka, S.R. and Sengupta, A. (2001). *Topics in Circular Statistics*, Volume 5. World Scientific, Singapore.

Johnson, D.S., London, J.M., Lea, M.-A., Durban, J.W. (2008). Continuous-time correlated random walk model for animal telemetry data. *Ecology*, 89(5), 1208–1215.

Lopez, R., Malardé, J.-P., Danès, P., Gaspar, P. (2015). Improving Argos Doppler location using multiple-model smoothing. *Animal Biotelemetry*, 3(1), 32.

Lunn, D.J., Thomas, A., Best, N., Spiegelhalter, D. (2000). Winbugs – A Bayesian modelling framework: Concepts, structure, and extensibility. *Statistics and Computing*, 10(4), 325–337.

Michelot, T., Langrock, R., Patterson, T.A. (2016). moveHMM: An R package for the statistical modelling of animal movement data using hidden Markov models. *Methods in Ecology and Evolution*, 7(11), 1308–1315 [Online]. Available at: https://besjournals.onlinelibrary.wiley.com/doi/abs/10.1111/2041-210X.12578.

Morales, J.M., Haydon, D.T., Frair, J., Holsinger, K.E., Fryxell, J.M. (2004). Extracting more out of relocation data: Building movement models as mixtures of random walks. *Ecology*, 85(9), 2436–2445.

Nathan, R., Getz, W.M., Revilla, E., Holyoak, M., Kadmon, R., Saltz, D., Smouse, P.E. (2008). A movement ecology paradigm for unifying organismal movement research. *Proceedings of the National Academy of Sciences*, 105(49), 19052–19059.

Patterson, T.A., Thomas, L., Wilcox, C., Ovaskainen, O., Matthiopoulos, J. (2008). State–space models of individual animal movement. *Trends in Ecology & Evolution*, 23(2), 87–94 [Online]. Available at: http://www.sciencedirect.com/science/article/pii/S0169534707003588.

Patterson, T.A., McConnell, B.J., Fedak, M.A., Bravington, M.V., Hindell, M.A. (2010). Using GPS data to evaluate the accuracy of state–space methods for correction of Argos satellite telemetry error. *Ecology*, 91(1), 273–285.

Rabiner, L.R. (1989). A tutorial on hidden Markov models and selected applications in speech recognition. *Proceedings of the IEEE*, 77(2), 257–286.

Tusell, F. (2011). Kalman filtering in R. *Journal of Statistical Software*, 39(2), 1–27.

de Valpine, P., Turek, D., Paciorek, C., Anderson-Bergman, C., Temple Lang, D., Bodik, R. (2017). Programming with models: Writing statistical algorithms for general model structures with NIMBLE. *Journal of Computational and Graphical Statistics*, 26, 403–413.

2

Detection of Eco-Evolutionary Processes in the Wild: Evolutionary Trade-Offs Between Life History Traits

Valentin JOURNÉ[1], Sarah CUBAYNES[2], Julien PAPAÏX[3] and Mathieu BUORO[4]

[1] *Grenoble Alpes University, INRAE, LESSEM, Saint-Martin-d'Hères, France*

[2] *CEFE, University of Montpellier, CNRS, EPHE-PSL University, IRD, Paul Valéry Montpellier 3 University, France*

[3] *Biostatistique et Processus Spatiaux (BioSP), INRAE, Avignon, France*

[4] *ECOBIOP, INRAE, Saint-Pée-sur-Nivelle, France*

2.1. Context

The biological lifecycle of organisms is characterized by a set of demographic traits, known as life history traits (e.g. size, growth, age at maturity, lifespan, etc.), which are often inter-related. In cases where these life history traits are positively dependent on a single limited resource, they exhibit negative correlation. If traits affect fitness components, in terms of survival and/or reproduction, then the dependency relationship is known as an evolutionary trade-off (Stearns 1989; Roff 2002; Flatt and Heyland 2011). If life history traits were independent, then individuals would simply seek to optimize each trait in order to maximize their own fitness. In reality, however, resources (time, space, energy, etc.) are limited, and must be shared, at individual level, between different traits that are essential to survival and reproduction (Metcalf 2016). While there may be environmental or genetic bases for

Statistical Models for Hidden Variables in Ecology,
coordinated by Nathalie PEYRARD and Olivier GIMENEZ.

trade-off (e.g. pleiotropic effects of genes), differential resource allocation is a widely cited explanation (Flatt and Heyland 2011; Descamps *et al.* 2016). For example, the allocation of resources to traits associated with reproduction at the present time may have a detrimental effect in terms of future survival or reproductive perspectives; this is known as the cost of reproduction (Williams 1966). Evolutionary trade-offs are considered as one of the most critical factors in the evolution of life history traits, and play a key role in the evolution of organisms, as they limit the range of possible adaptive variations (Stearns 1989; Flatt and Heyland 2011). One major challenge for evolutionary biologists is to detect and observe these trade-offs in natural environments, in order to understand and anticipate the evolution of life history traits (Metcalf 2016).

2.2. The correlative approach to detecting evolutionary trade-offs in natural settings: problems

While the existence of evolutionary trade-offs is widely accepted, their detection, along with the explanation of the underlying mechanisms, continues to present a major challenge in evolutionary biology. The study of evolutionary trade-offs was long limited by the lack of appropriate methods and by the existence of a number of confounding factors, due to the need to observe relevant life history traits and the environmental context (e.g. resources), something which is difficult to do in the natural environment.

Experimental studies can provide direct evidence of trade-offs by monitoring inter-individual variability, a potential source of confusion (e.g. Svensson *et al.* 2002; Bennett and Lenski 2007). This type of monitoring is not often possible in natural environments, making it much harder to study trade-offs. For this reason, a correlative approach between traits within a population is often used. Many empirical studies expected to reveal evolutionary trade-offs in natural environments have found positive (or null) correlations between life history traits where a negative correlation (i.e. trade-off) was expected (Bleu *et al.* 2016; King *et al.* 2011). This may result from high inter-individual variations in resource acquisition capacities, which may mask underlying trade-offs (Hamel *et al.* 2009; Metcalf 2016).

A trade-off may be seen as the result of (1) the total quantity of available resources, which depends on environmental conditions (mainly food supplies) and resource acquisition capacities, and (2) the strategy for resource allocation between two traits (van Noordwijk and de Jong 1986). Acquisition capacities and resource allocation strategies vary as a function of the environment, which determines the quantity of available resources, but also as a function of individual characteristics (van Noordwijk and de Jong 1986). Different traits may be maximized simultaneously in cases where resources are abundant, and in this case trade-offs may be masked. In cases where resources are limited, however, individuals with a greater capacity to acquire resources will be at an advantage, partly or totally avoiding the cost to which other individuals may be subject. If resources are limited, they will be allocated to one or more priority traits. Thus, a changing environment will modify both traits and co-variations between traits, which may be direct (e.g. a trade-off between growth and survival) or deferred (e.g. a trade-off between present and future reproduction). We thus expect to observe complex patterns of change in life history traits in the context of environmental variations.

Van Noordwijk and de Jong (1986) illustrate this concept using a "Y" model for cases where two traits (such as reproduction and survival) compete for the same resource at individual level, and where individuals differ both in terms of their capacity to acquire a resource (e.g. energy) and in the allocation of this resource to the traits in question. In this case, the quantity of a resource available for each characteristic is positively dependent on the total quantity of resources acquired and negatively dependent on the proportion of resources allocated to the other characteristic. In cases where inter-individual variations in resource acquisition is high relative to variation in allocation between traits, trade-offs at individual level are likely to go undetected in cases where their identification is based on inter-individual comparison (Figure 2.1). As certain individuals have a higher resource acquisition capacity than others (Hamel *et al.* 2009; Brown 2003), they have a greater ability to invest in both traits, and trade-offs will be masked. Thus, a true trade-off at individual level, based on limited resource allocation, may even translate to a positive correlation between traits on the basis of inter-individual covariation. This type of configuration is particularly frequent in natural conditions, where environmental variability is high and cannot be controlled by analysts, and where life histories are only partially observed.

2.2.1. *Mechanistic and statistical modeling as a means of accessing hidden variables*

Given the problems described above, trade-off analysis should be carried out using approaches that explicitly take account of the proximal mechanisms responsible for trade-offs (resource acquisition/allocation), and should focus on the individual level in order to take account of inter-individual resource acquisition and allocation strategies. This type of approach allows us to identify trade-offs in situations where identification is problematic due to natural variations. Nevertheless, proximal mechanisms are rarely observed directly in a natural environment; their hidden character means that evolutionary trade-offs remain hard to study.

In this chapter, we shall present two approaches based on both mechanistic and statistical models used to represent the expected proximal mechanisms and the conditioning structure which links life history traits, revealing the presence of underlying trade-offs. In the first example (section 2.3.1), the hidden proximal mechanisms are represented by latent variables, taking the form of theoretical quantities or variables that cannot be measured directly, but which may be supplied in the form of imperfect cues (observed variables). Trade-offs may thus be identified conditional on the status of the individuals for which variables have been observed. In other words, these approaches do not use data about resources themselves, but rather data concerning the results of individual acquisition and allocation of resources, known as proximal signals, which reflect energy sinks. In this example, a hierarchical Bayesian model is used to infer the existence of a trade-off. The second example (section 2.3.2) illustrates the use of another strategy based on the simulation of resources acquired at individual level. In this case, a mechanistic model, taking account of tree physiology, is used to obtain a value for the resources available to each individual. This value is then used as an input variable for a statistical model describing the distribution of the resource between different energy sinks.

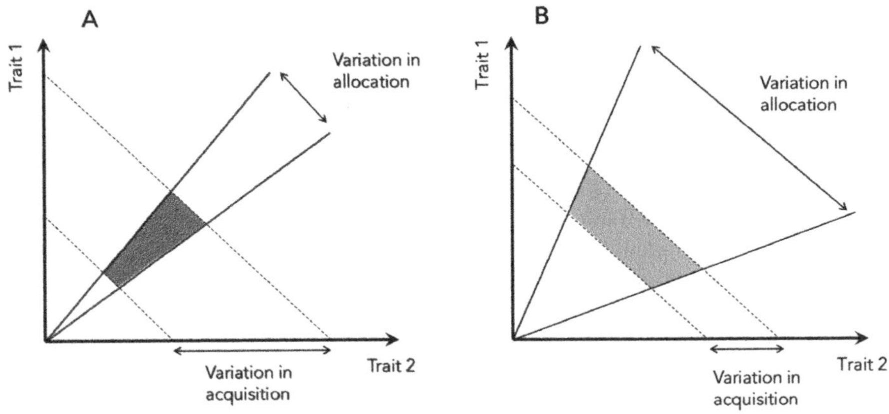

Figure 2.1. *Illustration of van Noordwijk and de Jong's (1986) "Y" model. Example of a trade-off between two traits 1 and 2, observed at population level (i.e. based on inter-individual comparison). A) In cases where the variation in resource acquisition is high in comparison with variations in allocation, a positive correlation between traits is observed at population level, masking the underlying trade-off (red). B) In cases where the variation in resource acquisition is low (limited resources) in relation to variations in allocation within the population, a negative correlation between traits is seen at population level, revealing the underlying trade-off (green). For a color version of this figure, see www.iste.co.uk/peyrard/ecology.zip*

2.3. Case study

2.3.1. *Costs of maturing and migration for survival: a theoretical approach*

We shall begin with a simple theoretical example of a trade-off between two life history tactics (e.g. reproduction and migration) in relation to a survival event. These three life history traits will be considered as the expression of latent processes, which are interdependent with regard to acquired resources (energy). The resources acquired by the individual, along with their allocation, cannot be observed directly, and are thus hidden states. Nevertheless, the expression of the life history traits resulting from this acquisition and from the differential allocation of resources can be observed.

2.3.1.1. *The system*

Our example is based on the lifecycle of the Atlantic salmon (*Salmo salar*), focusing on the juvenile phase, which is characterized by a sequence of events corresponding to (1) the decision to attain sexual maturity for the purposes of reproduction, (2) the decision to migrate, which is closely linked to growth, and (3) survival. Our approach is based on dynamic energy budget (DEB) theory (Sousa *et al.* 2010), which implies a division of the energy acquired by individuals (reserve) into structure and maturity (Kooijman *et al.* 2008). The justification for this distinction lies in the fact that maturity investment mostly takes the form of lipid storage, while structure is primarily dependent on the accumulation of protein resources (Sousa *et al.* 2010). Individual energy states will be defined as a function of these two bodily components. We shall also suppose that maturity storage is essentially used to fuel maturation processes (e.g. sexual development) and maintenance, while structural investment (i.e. growth) is essential for migration. Our hypothesis is that organisms are informed of their "state" (i.e. maturity and structure) by certain proximal signals (e.g. hormones). These proximal signals cannot be measured directly; they result from the acquisition and allocation of resources and are not observable. However, the mass of an individual may be used as an observable signal for stores, while size constitutes an observable signal for structure.

2.3.1.2. *Modeling life history decisions and proximal signals at the individual level*

Life history decisions: In our case, the two life history strategies and survival are observed binary variables, respectively $Y = (Y_1, Y_2, Y_3)'$. We have chosen to model life history decisions in terms of maturation and migration using a threshold modeling approach. We thus consider that individual decisions concerning maturation and migration result from a comparison between an intrinsic factor of the organism, represented by a proximal signal, and a threshold value. For an individual i, this implies that if the value of the proximal signal Z_i is higher (respectively, lower) than a threshold θ_i, then the expression of a phenotype, such as "mature" (respectively, immature) will be triggered. If Y_i is the binary indicator of the phenotype (e.g. taking 1 for mature and 0 for immature), then we have:

$$Y_i = \begin{pmatrix} 1 \ if \ Z_i > \theta_i \\ 0 \ if \ Z_i \leq \theta_i \end{pmatrix}$$

The proximal signal Z_i varies between individuals as a function of the environment, while the threshold θ_i also varies between individuals but is independent of the environment. The threshold is thus considered as an intrinsic property of individuals, which is independent of Z_i.

The proximal signal as a latent variable: Although the proximal signal Z_i is not observable, we may measure an observable signal X_i (e.g. size or mass) which correlates to Z_i. The distribution of the unknown proximal signal Z_i may be expressed in a manner conditional on the observable signal X_i with a certain residual error ϵ_i:

$$Z_i = F(X_i) + \epsilon_i$$

where F is a function, for example a linear relationship, summarizing the link between the proximal signal and the observable signal. The residual error ϵ_i is assumed normally distributed with a mean 0 and standard deviation σ_Z:

$$\epsilon_i \sim N(0, \sigma_Z)$$

This formulation has the advantage of being assumption-free with regard to the distribution of X_i. This means that statistical analysis is more flexible, as it is independent of the procedure used to collect observations X_i.

This formulation corresponds to the latent environmental threshold model (LETM) used by Buoro *et al.* (2012) to study life history decisions using empirical data. One important difference here is that, for reasons of identifiability and with no loss of generality, we shall take the threshold as fixed across individuals, whereas this value varies between individuals in the LETM.

Returning to our example, each successive event (sexual maturation, migration and survival) is characterized by a proximal signal, respectively, $\mathbf{Z} = (Z_1, Z_2, Z_3)$, and an associated threshold $\mathbf{\Theta} = (\theta_1, \theta_2, \theta_3)$.

2.3.1.3. *Modeling evolutionary trade-offs*

Expected cost of maturation and migration with respect to survival: In our case, the processes connected with maturation and migration are not in direct competition, and may thus occur simultaneously. However, we suppose that sexual maturation and preparation for migration all require energy, a resource which is also required to ensure survival. We thus expect to observe evolutionary trade-offs. Without becoming overly specific, let us establish some hypotheses concerning the relationships between life history traits:

1) The decision to mature and survive both depend positively on the amount of reserve. Given that the maturation process and survival are interdependent with regard to the quantity of stored energy available, we expect to observe a *negative trade-off*, that is, a cost of reproduction for survival.

2) Survival is negatively dependent on the state of the *structure*. The decision to migrate requires high growth (structural increase) and has a negative effect on chances of survival. For example, large individuals have a higher maintenance cost than smaller individuals, that is, energy requirements increase as size increases.

Implementation of costs with regard to survival: The survival state of an individual $Z_{3,i}$ is thus dependent on the relation between the state of reserves $Z_{1,i}$ and the structure $Z_{2,i}$. Our model accounts for potential underlying trade-offs using coefficients that affect the survival state $Z_{3,i}$ based on earlier life history decisions (maturation and migration). A coefficient α adjusts the status of the reserves $Z_{2,i}$ if the individual has already reached maturity ($Y_{i,1} = 1$), while a coefficient β is used to adjust the survival status $Y_{i,3}$ if the individual decides to migrate ($Y_{i,2} = 1$). Thus, for a given individual *i*, the proximal signal associated with survival $Z_{i,3}$ is

$$Z_{3,i} = \frac{Z_{1,i}}{Z_{2,i}} \times (\alpha \times Y_{1,i}) \times (\beta \times Y_{2,i}),$$

where Z_1, Z_2 and Z_3 are, respectively, the proximal signals for the maturation, migration and survival processes; $Y_{i,1}$ and $Y_{i,2}$ are binary indicators of the decision to mature and the decision to migrate. A linear transformation may be applied, for example using a natural logarithm, for ease of use. Parameters α and β are coefficients indicating the proportion of the remaining state conditional on life history decisions, that is, reflecting the effects of maturation and migration, respectively, on survival. Using a natural logarithm, if these parameters have a value of 1, there is no trade-off; if the coefficients are less than 1, there is a negative trade-off, while coefficients greater than 1 reflect a positive effect. This formulation implies that, while there is no cost for maturation or reproduction ($Y_1 = 0$ or $\alpha = 1$), the higher the maturation status (i.e. the greater the store), the higher the probability of survival (at constant migration status, i.e. structure). Conversely, the higher the migration status, the lower the probability of survival. Individuals who migrate invest their energy in growth rather than storage. The conditional structure of the model can be illustrated using a directed acyclic graph (Figure 2.2).

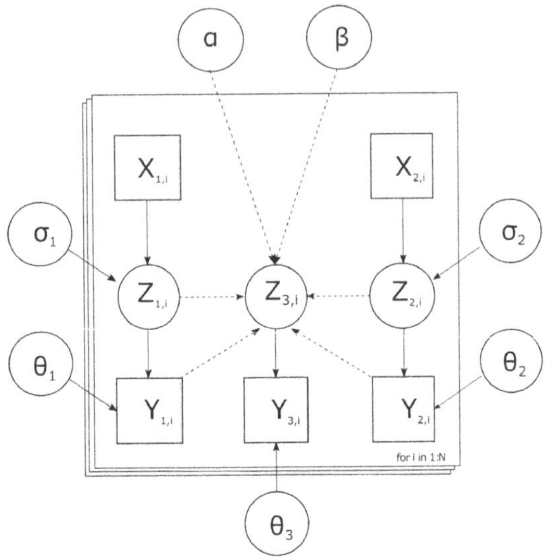

Figure 2.2. *Directed acyclic graph of the model. The squares represent observable data, while the circles show unknown quantities to estimate, i.e. latent variables. The plain and dashed arrows represent stochastic and determinist dependency, respectively. The model is designed for observations of phenotyped individuals hence the multiple frames, denoting a loop for i = 1, 2, ..., N individuals*

2.3.1.4. *Implementation and evaluation*

The Bayesian framework, using Markov chain Monte Carlo (MCMC) mechanisms, offers a flexible approach to the analysis of latent variable models and their conditional structure (Clark 2005). This approach is used here to estimate the parameters of interest in our model. Bayesian analysis requires the use of *a priori* probability distributions for the parameters of the model, that is, the unknown parameters (α, β, Θ and σs; Figure 2.2). In this case, all of the *a priori* distributions are non-informative. The joint *a posteriori* distributions for all of the unknowns in the model were obtained by MCMC sampling in the form used in JAGS, a program called up in R using the rjags packet (Plummer 2003). The convergence of the MCMC sampling was evaluated using the Brooks–Gelman–Rubin convergence diagnostic[1] (Brooks and Gelman 1998).

1 The code for the model and the R script used to generate data are available at: https://oliviergimenez.github.io/code_livre_variables_cachees/buoro.html.

2.3.1.5. *Identifying trade-offs*

The performance of our model for the statistical inference of parameters and in highlighting trade-offs is demonstrated here using simulated data based on observed life histories for juvenile Atlantic salmon. In order to ensure that our data are biologically realistic, we generated phenotypes based on the parameters of a length–weight relationship established using real data. These original data were collected from individuals via capture–recapture in the context of a long-term capture–recapture program for an Atlantic salmon population in the Scorff River (Morbihan, Brittany). We generated our simulated data using the model itself, with known parameter values (the fixed values are shown in Figure 2.3), considering three possible scenarios: (A) a negative trade-off ($\alpha = 0.3$ and $\beta = 0.8$), (B) no trade-off ($\alpha = 1$ and $\beta = 1$), and (C) a negative trade-off for maturation, but positive correlation for the decision to migrate ($\alpha = 0.3$ and $\beta = 1.2$). We generated 20 data sets of 500 individuals for each scenario. Statistical inference was then carried out using the simulated data to verify whether the model provided precise estimations of the parameters (α, β, Θ and σs).

For all scenarios, a comparison of *a posteriori* and *a priori* distributions shows that the *a priori* distributions were adjusted using the information contained in the data. The model provides a correct estimation of the residual variances σs and thresholds Θ; the *a posteriori* medians of these parameters were also close to their real values (Figure 2.3). The *a posteriori* distributions of α and β were correctly estimated, indicating that these parameters could be estimated, whatever the direction and scale of trade-off. Note that these parameters may be limited by identifiability issues, given the information available in the data. While we did not observe contrasting decisions in individual life histories (e.g. if all individuals mature, migrate and/or survive), the available information is necessarily based on the observed phenotypes alone.

In spite of the introduction of proximal mechanisms, our model remains relatively basic in the form presented here. It can be easily simplified (e.g. to focus on one life history strategy, using a linear model instead of the threshold model) or extended (e.g. with the addition of an observation process, a cost of maturing for migration, estimation of individual thresholds, etc.). Finally, while our simulated data draws on a simple energy allocation structure, more complex structures, such as a bioenergetic model, may be used. It is important to note that the analytical power of the model depends on the link between the hidden and observed variable, and, crucially, on the available data (e.g. in terms of the variability of life history strategies and traits).

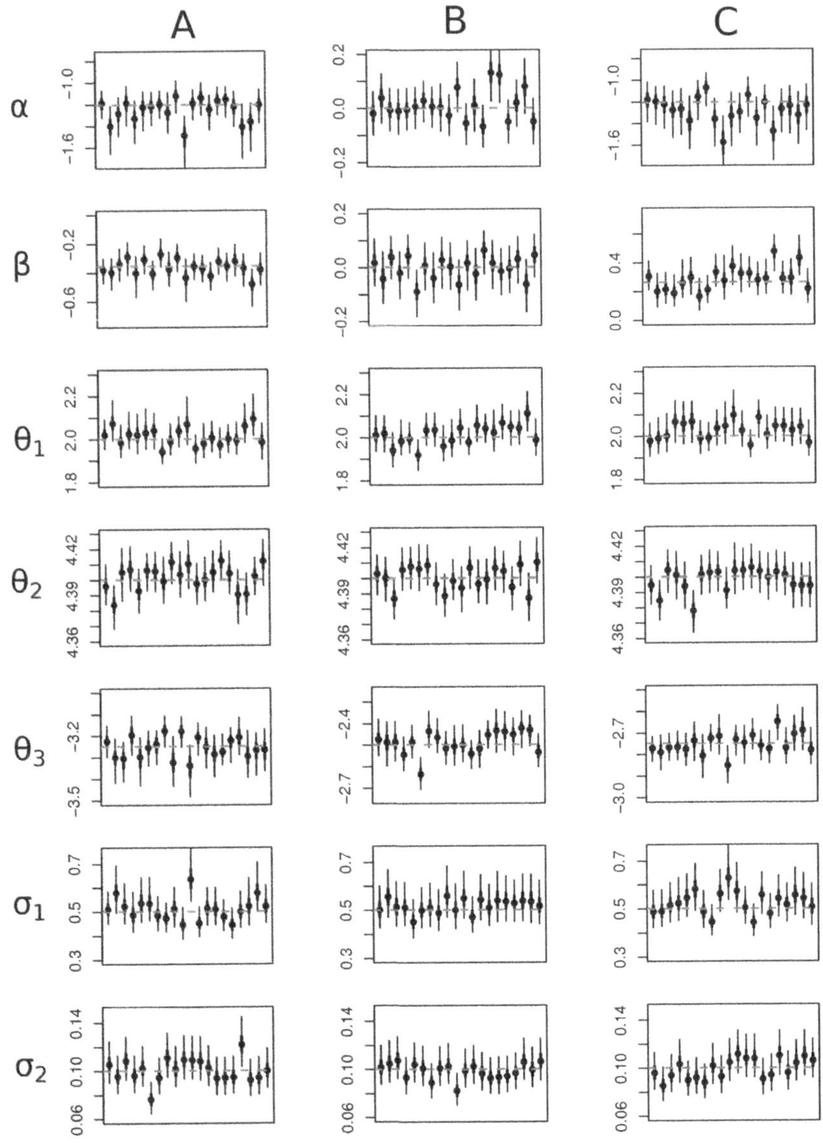

Figure 2.3. A posteriori *distributions of parameters in the latent model (logarithmic scale) for each scenario: A) negative trade-off, B) no trade-off and C) positive trade-off (migration only), for 20 simulated data sets. The median (black dot) and 95% confidence interval (continuous lines) were obtained on the basis of 25,000 MCMC iterations. Real values are shown as dashed red lines. For a color version of this figure, see www.iste.co.uk/peyrard/ecology.zip*

2.3.2. *Growth/reproduction trade-off in trees*

One key question linked to life history strategies concerns the expression of potential trade-offs between growth and reproduction, reflecting the fact that an individual cannot invest in both traits simultaneously. As we have seen, the main difficulty in studying these trade-offs within natural populations resides in the variability of individual resource acquisition. In this example, we illustrate the combination of a mechanistic eco-physiological model, used to simulate available resources at individual level, with a model representing the distribution of these resources between different energy sinks. Potential trade-offs are taken into account by means of correlated random effects between the different energy sinks. Inference is carried out using a Bayesian approach. The model is applied to data for the Atlas cedar (*Cedrus atlantica*), a Mediterranean conifer species[2].

2.3.2.1. *The studied system*

The Atlas cedar, *Cedrus atlantica*, is a coniferous species that originates from the Mediterranean basin. Individual trees carry both male and female cones. As in many tree species, male reproduction (from the initiation of the reproductive bud to the development of mature pollen) and growth occur over the course of a single year; female reproduction (from cone initiation to maturation) takes two years. The site of study is a 35-year experimental plantation in Mont-Ventoux in the south of France. All of the trees were planted and pedo-climatic conditions are similar across the site. The initial tree density was 2,700 trees per hectare. In this experiment, two different thinning strategies were applied, resulting in contrasting population densities at the point of observation: high density (1,200 trees per hectare) and low density (250 trees per hectare).

The data used here corresponds to 40 trees in the high-density population and 31 trees in the low-density population, randomly sampled and monitored each year from 2002 and 2005. The diameter of each individual tree i was measured at 1.3 m from the ground ($DBH_{i,t}$) for each year t. The annual basal area increment ($BAI_{i,t}^{\text{obs}}$) was calculated as $BAI_{i,t}^{\text{obs}} = (\pi \times DBH_t^2/4) - (\pi \times DBH_{t-1}^2/4)$. The abundance of male cones ($M_{i,t}$) was recorded as a qualitative ordered variable using a scoring system from 0 to 4, with "0" signifying that no male cones were observed; "1" few male cones dispersed in the canopy; "2" abundant male cones on one branch; "3" abundant male cones on two branches; "4" abundant male cones throughout the whole

2 The full article is available at: https://www.biorxiv.org/content/10.1101/2021.01.26.428205v1, while the data and the Bayesian model are available at: https://oliviergimenez.github.io/code_livre_variables_cachees/buoro.html.

of the tree canopy. These scores were then converted into multinomial observations $IMC_{i,t}^{\text{obs}}$ as follows:

$$\begin{cases} M_{i,t} = 0 \Rightarrow IMC_{i,t}^{\text{obs}} = [1, 0, 0, 0, 0] \\ M_{i,t} = 1 \Rightarrow IMC_{i,t}^{\text{obs}} = [0, 1, 0, 0, 0] \\ M_{i,t} = 2 \Rightarrow IMC_{i,t}^{\text{obs}} = [0, 0, 1, 0, 0] \\ M_{i,t} = 3 \Rightarrow IMC_{i,t}^{\text{obs}} = [0, 0, 0, 1, 0] \\ M_{i,t} = 4 \Rightarrow IMC_{i,t}^{\text{obs}} = [0, 0, 0, 0, 1] \end{cases} \quad [2.1]$$

Mature female cones ($FC_{i,t}^{\text{obs}}$) were counted individually from the ground, across the whole canopy, using binoculars.

2.3.2.2. *Resource simulation at individual level*

Simulations from the CASTANEA eco-physiological model were used to determine a level of resources, taking account of the characteristics of individual trees in terms of diameter, leaf area index, allometric relationship and population density (Davi *et al.* n.d.; Dufrêne *et al.* 2005; Davi and Cailleret 2017). Resources are represented by net primary productivity, NPP, which corresponds to the difference between gross photosynthesis and autotrophic respiration. CASTANEA takes account of forest characteristics (such as tree height and diameter, density, nitrogen and sugar concentrations), soil conditions (such as texture) and climate. Climate was assessed using local variables, measured from 1999 to 2005 at a weather station located 1.96 km from the study site. Additional data were obtained from the national meteorological database (SAFRAN) using measurements from 1989 to 2015, scaled using statistical regression to correspond to local measurements (Quintana-Seguí *et al.* 2008). The variables used by the CASTANEA model include precipitation; minimum, maximum and mean temperature; overall radiation; relative humidity; and wind speed. The model was validated for the study site and for both stand densities based on its capacity to simulate a mean annual growth value. The data and output for the CASTANEA model was evaluated (Figure 2.4) using mean quadratic error ($RMSE$), the determination coefficient (R^2) and percentage bias (PB). Figure 2.5 shows resource distribution between individuals simulated using CASTANEA. The effect of density on resource availability is clearly visible.

2.3.2.3. *Modeling evolutionary trade-offs*

Resource allocation to energy sinks: The resource level ($NPP_{i,t}$) for an individual i in year t determines its growth ($BAI_{i,t}$), the number of initiated reproductive buds ($IB_{i,t}$) and the probability of female cone survival ($p_{i,t}^{FCS}$). Random effects ($\epsilon_{x,i}$) were used to study potential trade-offs between the three sinks. Two alternative models were compared. In model 1, inter-individual variation acts directly on the quantity of resource available (i.e. corresponding to the slope). In model 2, inter-individual variation is directly connected to the trait (i.e. corresponding to the intercept). The model is defined as follows (see Figure 2.6):

$$\begin{cases} BAI_{i,t} = (\gamma_d + \underbrace{Y * \epsilon_{1,i})}_{\text{model 1}} * NPP_{i,t} + \underbrace{(1-Y) * \epsilon_{1,i}}_{\text{model 2}} \\ IB_{i,t} = X_{i,t} * ((\beta_{1,d} + \underbrace{Y * \epsilon_{2,i})}_{\text{model 1}} * NPP_{i,t} + \underbrace{(1-Y) * \epsilon_{2,i})}_{\text{model 2}} \\ \text{logit}(p_{i,t}^{FCS}) = \beta_0 + (\beta_{2,d} + \underbrace{Y * \epsilon_{3,i})}_{\text{model 1}} * NPP_{i,t} + \underbrace{(1-Y) * \epsilon_{3,i}}_{\text{model 2}} \\ \epsilon \sim \mathcal{N}_3(0, \sum). \end{cases} \quad [2.2]$$

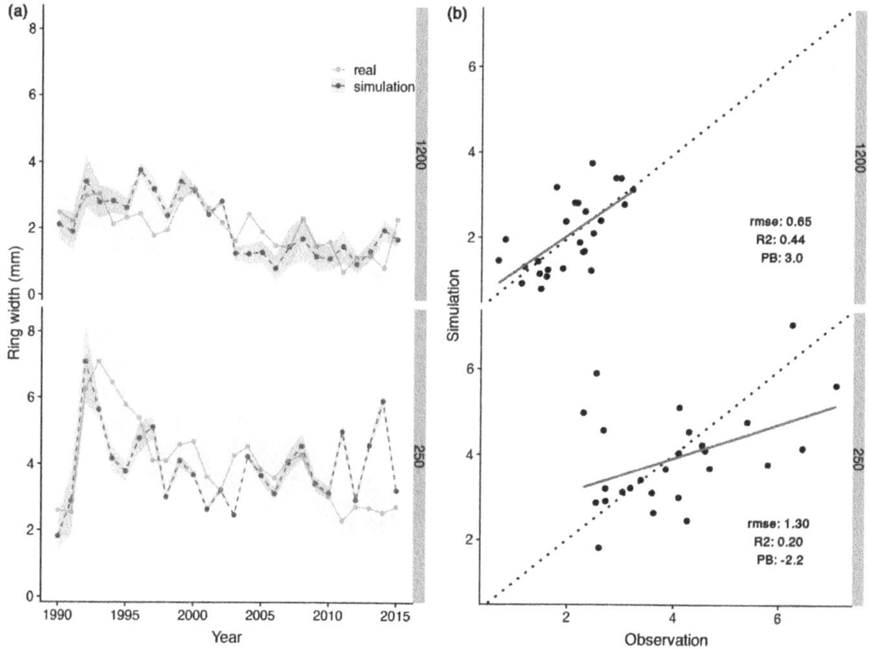

Figure 2.4. *Comparison of observed and simulated ring width from 1989 to 2015 and for two stand densities. a) Observed data are shown in light blue, while simulated ring widths are shown in dark blue. Mean values are shown along with standard deviations (broader strip). b) The dashed black line corresponds to the 1:1 line, and the red line corresponds to the regression between simulated and observed ring width values. For a color version of this figure, see www.iste.co.uk/peyrard/ecology.zip*

where the random variable $Y \sim$ Bernoulli (p_Y) is used to select model 1 or 2, based on its value, 0 or 1. Parameters γ_d, $\beta_{1,d}$ and $\beta_{2,d}$ are the slopes cconnecting resources to different sinks. The parameter β_0 is fixed to limit $p_{i,t}^{FCS} \approx 0$ when $NPP_{i,t} = 0$. The random variable $X_{i,t} \sim$ Bernoulli (p_X) indicates whether or not an individual i produces reproductive buds in year t. Resource allocation constraints for all three sinks are accounted for using individual effects $(\epsilon_{1,i}, \epsilon_{2,i}, \epsilon_{3,i})^T \sim$ Normal$_3$ $(0, \Sigma)$. Correlations are calculated as $\rho_{l,k} = \Sigma_{l,k} / (\Sigma_{l,l} \Sigma_{k,k})$.

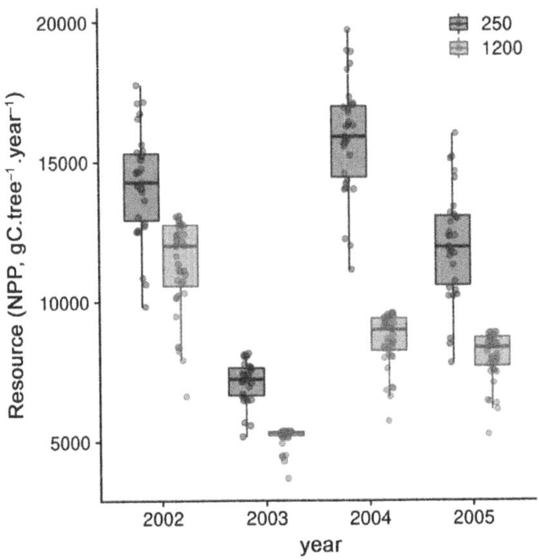

Figure 2.5. *Boxplot of resource (net primary productivity, in gC.m^{-2}.year^{-1}) simulated using the eco-physiological CASTANEA model. For a color version of this figure, see www.iste.co.uk/peyrard/ecology.zip*

Modeling reproduction: Initiated reproductive buds $(IB_{i,t})$ develop into male cones $(IMC_{i,t})$ and a number of initiated female cones $(IFC_{i,t})$ depending on the phenotypic gender $(PG_{i,t})$ of an individual i in year t. Phenotypic gender corresponds to "maleness" Lloyd 1980), which is the ratio of male to female initiated cones. Thus,

$$\begin{cases} \text{logit}\,(PG_{i,t}) \sim \mathcal{N}\left(\bar{PG}, \sigma_{PG}\right) \\ IMC_{i,t} = PG_{i,t} * IB_{i,t} \\ IFC_{i,t} = (1 - IMC_{i,t}) \, . \end{cases} \quad [2.3]$$

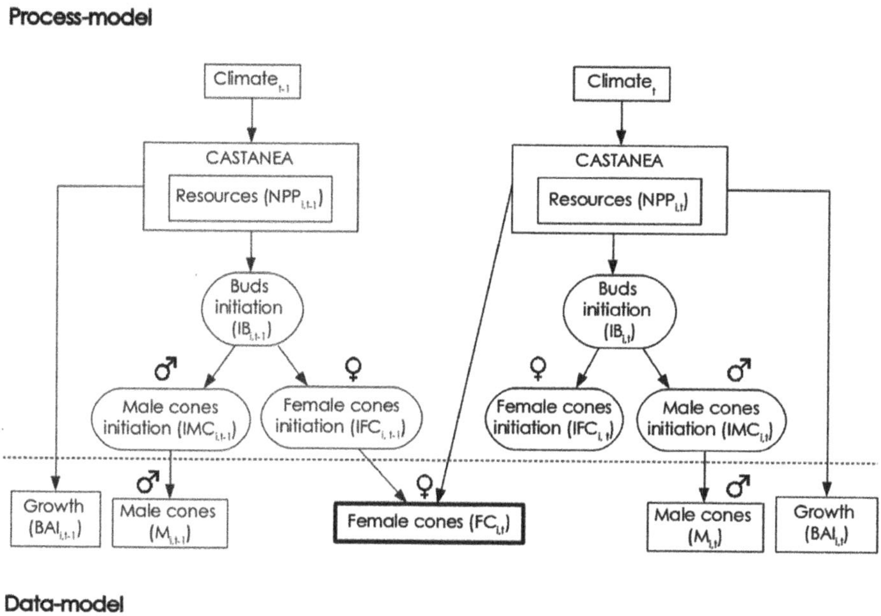

Figure 2.6. *Illustration of the developed Bayesian model, with process and data models. The rectangles correspond to observed variables and the ellipses to non-observed (i.e. latent) variables. The previous year ($t-1$) is shown in gray and the current year (t) in black. Resources, in terms of net primary productivity (NPP), are simulated using the CASTANEA eco-physiological model and climate data. Resources are then allocated to tree growth (Basal Area Increment, $BAI_{i,t}$) and to reproduction, initially represented by initiation buds, $IB_{i,t}$. Buds are then subdivided into male ($IMC_{i,t}$) and female cones ($IFC_{i,t}$) according to phenotypic gender (not shown, noted $PG_{i,t}$). A given year t will see the maturing of male cones ($M_{i,t}$) alongside that of female cones, which depends on the quantity of buds initiated in the previous year ($t-1$)*

Observation process: The available data (see 2.3.2.1) consists of repeated observations of growth, male reproduction and female reproduction. Observed growth ($BAI^{\text{obs}}_{i,t}$) is considered to be linked to the latent growth variable of the process model ($BAI_{i,t}$) as follows:

$$BAI^{\text{obs}}_{i,t} \sim \mathcal{N}\left(BAI_{i,t}, \sigma_{BAI}\right). \quad [2.4]$$

The number of initiated male cones ($IMC_{i,t}$) is a continuous variable, while the observed abundance of male cones ($IMC_{i,t}^{\text{obs}}$) in our example is a categorical variable. $IMC_{i,t}$ and $IMC_{i,t}^{\text{obs}}$ are linked here using the following model:

$$\begin{cases} \boldsymbol{\pi_{i,t}} = [F(s_0), F(s_1) - F(s_0), F(s_2) - F(s_1), F(s_3) - F(s_2), 1 - F(s_3)] \\ IMC_{i,t}^{\text{obs}} \sim Multinomial\,(\boldsymbol{\pi_{i,t}}, 1)\,, \end{cases} \quad [2.5]$$

where $F(\cdot)$ indicates the distribution function of a normal distribution of mean $IMC_{i,t}$ and variance σ_{IMC}. $\{s_0, s_1, s_2, s_3\}$ are fixed thresholds that determine the limit between male bud scores. Finally, the observed number of female cones ($FC_{i,t}^{\text{obs}}$) follows a Poisson distribution of parameter $IFC_{i,t}$, in which the latent variable denotes the number of initiated female cones:

$$FC_{i,t}^{\text{obs}} \sim \mathcal{P}(p_{i,t}^{FCS} * IFC_{i,t-1} * IB_{i,t-1}). \quad [2.6]$$

2.3.2.4. *Implementation and fitting*

Parameters were estimated using JAGS (Plummer 2003) (version 4.3.0) in R (R Core Team 2018) (version 3.6.3). Non-informative prior distributions were defined for all parameters. Three MCMC of 200,000 iterations were simulated. The fitting of the model to the data was evaluated by calculating the Bayesian p-value for quantitative variables (Gelman *et al.* 1996) and using the Brier score for the qualitative variable $IMC_{i,t}^{\text{obs}}$.

Model 2, in which individual effects are not linked to resources, performed best. Thus, it appears that NPP is not the only factor responsible for correlations between energy sinks. The NPP may be involved in an indirect manner via a density effect, whereby individual NPP is lower in high-density stands, but other factors, such as the availability of nitrogen, minerals or other energy sources, seem to contribute to correlations.

2.3.2.5. *Trade-off identification*

The estimated correlations between individual effects associated with each energy sinks (Figure 2.7) show that individuals exhibiting high growth rates also produce large numbers of reproductive buds ($\rho_{1,2} = 0.54[0.32, 0.70]$ with $pr[\rho_{1,2} > 0] = 1$). There is thus no visible trade-off between the two functions. However, a trade-off is seen between the number of initiated buds and the survival of female cones from 1 year to the next ($\rho_{1,3} = -0.36[-0.65, 0.24]$ with $pr[\rho_{1,3} < 0] = 0.91$). Thus, trees which invest more resources in maturing cones produce fewer reproductive buds. This relationship has been widely observed in a variety of plant species (Bell 1980; Knops *et al.* 2007).

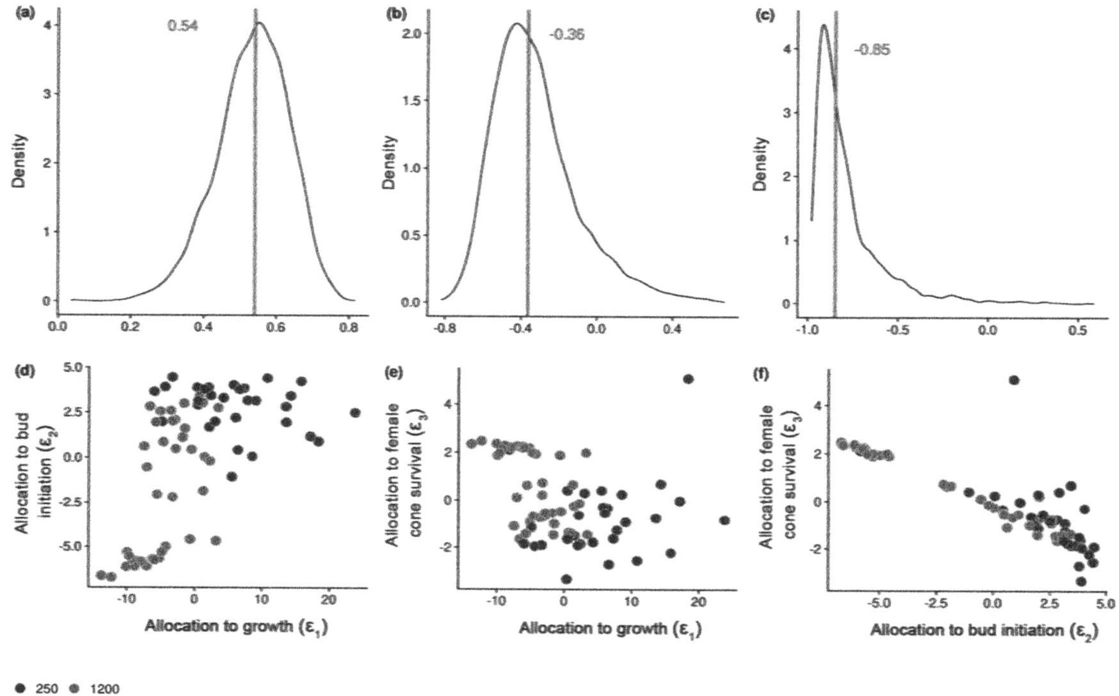

Figure 2.7. *Correlation between sinks and probabilities. a) Correlation density $\rho_{1,2}$ is positive. b) Correlation density $\rho_{1,3}$ is negative. c) Correlation density $\rho_{1,2}$ is negative. The red line shows the median value of the posterior in each figure. d) Positive correlation between growth allocation (ϵ_1) and allocation to reproductive bud initiation (ϵ_2). e) Negative correlation between growth allocation (ϵ_1) and the survival of female cones (ϵ_3). f) Negative correlation between initiation of reproductive buds (ϵ_2) and the survival of female cones (ϵ_3). The points show mean individual values for each of the two population densities. For a color version of this figure, see www.iste.co.uk/peyrard/ecology.zip*

Trees in the low density stand generally allocate more resources to growth than to reproduction ($pr[\gamma_{1200} > \gamma_{250}] = 0.25$), unlike individuals in the high-density stand ($pr[\beta_{1,1200} > \beta_{1,250}] = 0.86$ and $pr[\beta_{2,1200} > \beta_{2,250}] = 0.94$). Furthermore, the correlations between sinks in terms of resource allocation are also dependent on density; these relationships are more clearly visible in the high-density stand, where resource levels are lower.

2.4. References

Bell, G. (1980). The costs of reproduction and their consequences. *The American Naturalist*, 116(1), 45–76 [Online]. Available at: http://www.jstor.org/stable/2460709.

Bennett, A.F. and Lenski, R.E. (2007). An experimental test of evolutionary trade-offs during temperature adaptation. 104(1), 8649–8654.

Bleu, J., Gamelon, M., Sæther, B.-E. (2016). Reproductive costs in terrestrial male vertebrates: Insights from bird studies. *Proceedings of the Royal Society B: Biological Sciences*, 283(1823), 20152600 [Online]. Available at: https://royalsocietypublishing.org/doi/abs/10.1098/rspb.2015.2600.

Brooks, G.L and Gelman, A. (1998). General methods of monitoring convergence of iterative simulations. *Journal of Computational ad Graphical Statistics*, 7(4), 434–455.

Brown, C.A. (2003). Offspring size-number trade-offs in scorpions: An empirical test of the van Noordwijk and de Jong model. *Evolution*, 57(9), 2184–2190 [Online]. Available at: https://onlinelibrary.wiley.com/doi/abs/10.1111/j.0014-3820.2003.tb00397.x.

Buoro, M., Gimenez, O., Prévost, E. (2012). Assessing adaptive phenotypic plasticity by means of conditional strategies from empirical data: The latent environmental threshold model. *Evolution*, 66(4), 996–1009 [Online]. Available at: https://onlinelibrary.wiley.com/doi/abs/10.1111/j.1558-5646.2011.01484.x.

Clark, J.S. (2005). Why environmental scientists are becoming Bayesians. *Ecology Letters*, 8(1), 2–14 [Online]. Available at: https://onlinelibrary.wiley.com/doi/abs/10.1111/j.1461-0248.2004.00702.x.

Davi, H. and Cailleret, M. (2017). Assessing drought-driven mortality trees with physiological process-based models. *Agricultural and Forest Meteorology*, 232, 279–290.

Davi, H., Barbaroux, C., Francois, C., Dufrêne, E. (n.d.). The fundamental role of reserves and hydraulic constraints in predicting LAI and carbon allocation in forests. *Agricultural and Forest Meteorology*, (2), 349–361.

Descamps, S., Gaillard, J.-M., Hamel, S., Yoccoz, N. (2016). When relative allocation depends on total resource acquisition: Implication for the analysis of trade-offs. *Journal of Evolutionary Biology*, 29(9), 1860–1866.

Dufrêne, E., Davi, H., François, C., Le Maire, G., Le Dantec, V., Granier, A. (2005). Modelling carbon and water cycles in a beech forest. Part I: Model description and uncertainty analysis on modelled NEE. *Ecological Modelling*, 185(2–4), 407–436.

Flatt, T. and Heyland, A. (2011). *Mechanisms of Life History Evolution: The Genetics and Physiology of Life History Traits and Trade-offs*. OUP, Oxford.

Gelman, A., Meng, X.-L., Stern, H. (1996). Posterior predictive assessment of model fitness via realized discrepancies. *Statistica Sinica*, 6(4), 733–760.

Hamel, S., Côté, S.D., Gaillard, J.-M., Festa-Bianchet, M. (2009). Individual variation in reproductive costs of reproduction: High-quality females always do better. *Journal of Animal Ecology*, 78(1), 143–151.

King, E., Roff, D., Fairbairn, D. (2011). Trade-off acquisition and allocation in Gryllus firmus: A test of the Y model. *Journal of Evolutionary Biology*, 24(2), 256–264.

Knops, J.M.H., Koenig, W.D., Carmen, W.J. (2007). Negative correlation does not imply a tradeoff between growth and reproduction in California oaks. *Proceedings of the National Academy of Sciences*, 104(43), 16982–16985 [Online]. Available at: http://www.pnas.org/cgi/doi/10.1073/pnas.0704251104.

Kooijman, S., Sousa, T., Pecquerie, L., Van der Meer, J., Jager, T. (2008). From food-dependent statistics to metabolic parameters, a practical guide to the use of dynamic energy budget theory. *Biological Reviews*, 83(4), 533–552.

Lloyd, D.G. (1980). Sexual strategies in plants III. A quantitative method for describing the gender of plants. *New Zealand Journal of Botany*, 18, 103–108.

Metcalf, C.J.E. (2016). Invisible trade-offs: van Noordwijk and de Jong and life-history evolution. *The American Naturalist*, 187(4), iii–v [Online]. Available at: https://doi.org/10.1086/685487.

van Noordwijk, A.J. and de Jong, G. (1986). Acquisition and allocation of resources: Their influence on variation in life history tactics. *The American Naturalist*, 128, 137–142.

Plummer, M. (2003). JAGS: A program for analysis of Bayesian graphical models using Gibbs sampling. *Proceedings of the 3rd International Workshop on Distributed Statistical Computing*, Vienna, March 20–22.

Quintana-Seguí, P., Le Moigne, P., Durand, Y., Martin, E., Habets, F., Baillon, M., Canellas, C., Franchisteguy, L., Morel, S. (2008). Analysis of near-surface atmospheric variables: Validation of the SAFRAN analysis over France. *Journal of Applied Meteorology and Climatology*, 47(1), 92–107 [Online]. Available at: http://journals.ametsoc.org/doi/abs/10.1175/2007JAMC1636.1.

R Core Team (2018). *R: A language and environment for statistical computing*. R Foundation for Statistical Computing, Vienna, Austria [Online]. Available at: https://www.R-project.org/.

Roff, D. (2002). *Life History Evolution*. Sinauer Associates, inc., Publishers Sunderland, MA.

Sousa, T., Domingos, T., Poggiale, J.-C., Kooijman, S. A. L.M. (2010), Dynamic energy budget theory restores coherence in biology. *Philosophical Transactions of the Royal Society B: Biological Sciences*, 365(1557), 3413–3428 [Online]. Available at: https://royalsocietypublishing.org/doi/abs/10.1098/rstb.2010.0166.

Stearns, S.C. (1989). Trade-offs in life-history evolution. *Functional Ecology*, 3(3), 259–268.

Svensson, E., Sinervo, B., Comendant, T. (2002). Mechanistic and experimental analysis of condition and reproduction in a polymorphic lizard. *Journal of Evolutionary Biology*, 15(6), 1034–1047 [Online]. Available at: https://www.onlinelibrary.wiley.com/doi/abs/10.1046/j.1420-9101.2002.00452.x.

Williams, G.C. (1966). Natural selection, the costs of reproduction, and a refinement of Lack's principle. *The American Naturalist*, 100(916), 687–690.

3

Studying Species Demography and Distribution in Natural Conditions: Hidden Markov Models

Olivier GIMENEZ[1], Julie LOUVRIER[2], Valentin LAURET[1] and Nina SANTOSTASI[3]

[1] *CEFE, University of Montpellier, CNRS, EPHE, IRD, Paul Valéry Montpellier 3 University, France*

[2] *Department of Ecological Dynamics, Leibniz Institute for Zoo and Wildlife Research, Berlin, Germany*

[3] *Department of Biology and Biotechnologies "Charles Darwin", University of Rome La Sapienza, Italy*

3.1. Introduction

Ecology may be defined as the study of living organisms in interaction with their environment. At the heart of this discipline lie two key questions: how many individuals are there in a population and where are they? The first question relates to the dynamics of populations, while the second concerns the distribution of species. These questions have long attracted the interest of researchers; for example, in the early 19th century, Laplace attempted to estimate the size of the French population (Amorös 2014), while at the start of the 20th century, Grinnell (1917) focused on formalizing the role of species in ecosystem functioning.

Statistical research in relation to these questions continues to this day, notably motivated by of the analysis of data generated using new technologies (Gimenez *et al.*

Statistical Models for Hidden Variables in Ecology,
coordinated by Nathalie PEYRARD and Olivier GIMENEZ.

2014b). One issue that has attracted particular attention is the difficulty of observing individuals and species in natural conditions – essentially, a detection problem (Royle and Dorazio 2008). Given the imperfections inherent in the detection of individuals and species, variables such as whether an individual is dead or alive, or whether or not a species is present in a particular location, are only partially observable; as such, they constitute hidden variables, in the sense defined in the introduction to this book.

In this chapter, we shall show how hidden Markov models (HMMs) can be used to develop capture–recapture and occupancy models, traditionally used to study the dynamics of populations and the distribution of species in a context of imperfect detection. We shall show how the HMM formulation permits the estimation of hidden variables in two different case studies. The question of population dynamics will be illustrated through an estimation of the prevalence of wolf-dog hybrids in Italy, while the distribution of species will be illustrated by examining the distribution of wolves in France.

3.2. Overview of HMMs

HMMs are a class of statistical models, generally used for analyzing data from systems with temporal dynamics. An ecological process may be modeled using a state process (or system process) of which the future states are solely dependent on current states: this is the Markov hypothesis. In an HMM, this process is not observed directly, but is hidden (latent). Observations are made based on a state-dependent process, controlled by the underlying state process. These observations are essentially considered to be noisy measures of system states with a specific dependence structure. HMMs are a specific class of state space models with a finite number of states (Gimenez *et al.* 2012; Auger-Méthé *et al.* 2020).

In formal terms, an HMM consists of an observed state-dependent process $Y^1, Y^2, \dots, Y^T$ and a non-observed (hidden) state process $Z^1, Z^2, \dots, Z^T$. HMMs are often represented schematically in the way shown in Figure 3.1, which highlights the way in which observations are conditional on states, and illustrates the Markovian structure of the sequence of states.

Three components are needed to fully specify an HMM with N states. The first component is the initial distribution $\boldsymbol{\delta} = \big(\Pr(Z^1 = 1), \dots, \Pr(Z^1 = N)\big)$, which combines the probabilities of being in different states at the start of the sequence. The second component is made up of the probabilities of transition $\gamma_{ij} = \Pr(Z^{t+1} = j | Z^t = i)$ between states i and j, generally grouped into an $N \times N$ transmission matrix, $\boldsymbol{\Gamma} = (\gamma_{ij})$. The third component is the distribution of the collected observations $f(y^t | Z^t = i)$, used to facilitate the calculation of likelihood in a diagonal matrix of dimension $N \times N$, denoted as $\mathbf{P}(y^t) = \text{diag}\big(f(y^t|Z^t = 1), \dots, f(y^t|Z^t = N)\big)$. In this chapter, only discrete and univariate distributions of observations will be addressed, but continuous distributions (Choquet *et al.* 2017; Mews *et al.* 2020) and multivariate distributions (Choquet *et al.* 2013; Laake *et al.* 2014; Johnson *et al.* 2016) may also be used.

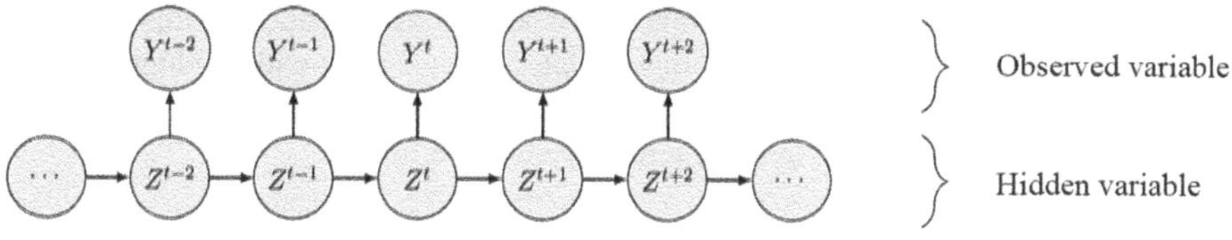

Figure 3.1. *Schematic illustration of a hidden Markov model*

The likelihood $\mathcal{L}(\boldsymbol{\theta} \mid y^1, \ldots, y^T)$ of the unknown parameters ($\boldsymbol{\theta}$) given an observed sequence $(Y^1, \ldots, Y^T)$ is expressed formally as:

$$\begin{aligned}\mathcal{L}(\boldsymbol{\theta} \mid y^1, \ldots, y^T) &= f_{\boldsymbol{\theta}}(y^1, \ldots, y^T) \\ &= \sum_{z^1=1}^{N} \cdots \sum_{z^T=1}^{N} f_{\boldsymbol{\theta}}(y^1, \ldots, y^T | z^1, \ldots, z^T) f_{\boldsymbol{\theta}}(z^1, \ldots, z^T) \\ &= \sum_{z^1=1}^{N} \cdots \sum_{z^T=1}^{N} \delta_{z^1} \prod_{t=1}^{T} f_{\boldsymbol{\theta}}(y^t | z^t) \prod_{t=2}^{T} \gamma_{z^{t-1}, z^t}.\end{aligned}$$

The first step is obtained by applying the law of total probability; the second step is a result of the Markovian dependency structure of the model.

The problem lies in the fact that the calculation of the likelihood of an HMM in this form requires N^T summations, making it time consuming, if not impossible, to evaluate. One solution is to use a more efficient method to calculate the likelihood. In this chapter, we have chosen to use the forward algorithm, which draws on the dependency structure of the model, instead of a "brute force" approach that consists of summing all possible series of states.

Using the forward algorithm, likelihood is calculated as a matrix product:

$$\mathcal{L}(\boldsymbol{\theta} \mid y^1, \ldots, y^T) = \boldsymbol{\delta} \mathbf{P}(y^1) \boldsymbol{\Gamma} \mathbf{P}(y^2) \cdots \boldsymbol{\Gamma} \mathbf{P}(y^{T-1}) \boldsymbol{\Gamma} \mathbf{P}(y^T) \mathbf{1},$$

where $\mathbf{1}$ is a column vector of ones. The complexity of this calculation is linear as a function of the number of observations, meaning that likelihood can be evaluated rapidly in most of the cases encountered in ecology. The parameters $\boldsymbol{\theta}$ of an HMM can be calculated by maximum likelihood, using optimization routines (such as the

Newton–Raphson method) to maximize likelihood numerically. This is the approach used here, implemented using R.

Once the parameters have been estimated, the next step is to infer the hidden states $z^1, \ldots, z^T$. In the context of HMM, this step is known as decoding. In this case, we use global decoding to look for the series of states $(g^1, \ldots, g^T)$ with the highest joint probability (this differs from local decoding, in which we search for the most likely value of z^t taken separately). In other terms, we wish to find

$$(g^1, \ldots, g^T) = \underset{(z^1,\ldots,z^T)}{\arg\max}\ \Pr(Z^1 = z^1, \ldots, Z^T = z^T \mid y^1, \ldots, y^T).$$

This is a relatively complex optimization problem; however, it can be solved efficiently using the Viterbi algorithm (Rabiner 1989).

For more details on HMMs in general, see Zucchini *et al.* (2016); in the ecological context, see McClintock *et al.* (2020).

3.3. HMM and demography

3.3.1. *General overview*

The hidden (or partially hidden) variables encountered in the study of animal populations are living/dead; developmental states, which are generally discrete, such as sexual maturity (Nichols *et al.* 1994); epidemiological states (Marescot *et al.* 2018); or social states (Dupont *et al.* 2015). These states can be hard to measure in the field. It is often impossible to track animals in their environment in an exhaustive manner, that is, in the way human patients might be monitored in the context of a medical protocol. Data are often obtained in capture–recapture form, indicating whether or not an animal has been detected. If an individual is not detected, it may be possible to infer its state; if an individual is detected, then its state may be known perfectly or imperfectly. HMMs are a natural choice for use in these contexts, as they can be used to formalize the analysis of noisy measures of demographic states.

One example involves the two states “dead” and “alive”, with $Z^t = \mathrm{V}$ denoting “alive at time t” and $Z^t = \mathrm{M}$ “dead at time t” (Gimenez *et al.* 2007). The “dead” state here is absorbing as an individual cannot leave the state once it has entered it (except in the context of zombie movies). An illustration of the corresponding HMM is given in Figure 3.2.

As we have seen, an HMM is defined using three components. The initial distribution is:

$$\delta = \begin{array}{c} \begin{matrix} \mathrm{V} & \mathrm{M} \end{matrix} \\ \begin{pmatrix} 1 & 0 \end{pmatrix} \end{array}.$$

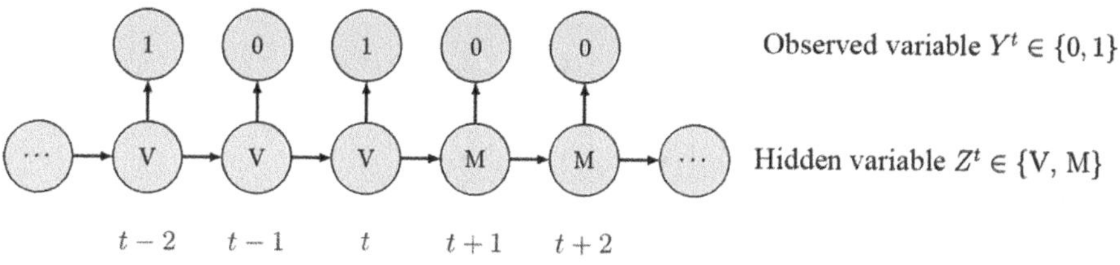

Figure 3.2. *Two-state capture–recapture model expressed in HMM form*

Let ϕ be the probability of survival over an interval of time. The transition probability matrix is given by:

$$\mathbf{\Gamma} = \begin{array}{c} \begin{array}{cc} \text{V} & \text{M} \end{array} \\ \begin{bmatrix} \phi & 1-\phi \\ 0 & 1 \end{bmatrix} \end{array} \begin{array}{c} \text{V} \\ \text{M} \end{array}.$$

Finally, the distribution of observations Y^t conditional on the states Z^t is a Bernoulli distribution of parameter p, where p is the probability of detection, if $Z^t = V$, or a Bernoulli distribution of parameter 0 if $Z^t = M$:

$$\mathbf{P}(y^t) = \begin{array}{c} \begin{array}{cc} \text{V} & \text{M} \end{array} \\ \begin{bmatrix} p^{y^t}(1-p)^{1-y^t} & 0 \\ 0 & 1-y^t \end{bmatrix} \end{array}.$$

Thus, if the individual is dead, $Z^t = M$, then the probability of observation is null, $\Pr(y^t = 1|Z^t = M) = 1 - y^t = 0$, and the probability of it not being observed is 1, $\Pr(y^t = 0|Z^t = M) = 1 - y^t = 1$. If the individual is alive, $Z^t = V$, the probability of observing it is $\Pr(y^t = 1|Z^t = V) = p^{y^t}(1-p)^{1-y^t} = p$, and the probability of it not being observed $\Pr(y^t = 0|Z^t = V) = p^{y^t}(1-p)^{1-y^t} = 1 - p$.

The contribution of each individual to the overall likelihood of the data set can then be calculated using these components. For example, consider a study that takes place over the course of $T = 3$ years, and let us take an individual observed in the first and third years, but not in the second year: $(y^1 = 1, y^2 = 0, y^3 = 1)$. This individual's contribution to the likelihood is written as:

$$\mathcal{L}(\phi, p \mid y^1, y^2, y^3) = f_{\phi,p}(y^1, y^2, y^3) = \boldsymbol{\delta}\mathbf{P}(y^1)\mathbf{\Gamma}\mathbf{P}(y^2)\mathbf{\Gamma}\mathbf{P}(y^3)\mathbf{1},$$

with

$$\mathbf{P}(y^1) = \mathbf{P}(y^3) = \begin{array}{c} \begin{array}{cc} \text{V} & \text{M} \end{array} \\ \begin{bmatrix} p & 0 \\ 0 & 0 \end{bmatrix} \end{array}$$

and

$$\mathbf{P}(y^2) = \begin{array}{c} \begin{array}{cc} \text{V} & \text{M} \end{array} \\ \begin{bmatrix} 1-p & 0 \\ 0 & 1 \end{bmatrix} \end{array}.$$

We can verify (with a little patience) that this matrix product is equal to $p\phi(1-p)\phi p$, generally conditioned with respect to the first capture, with an assigned value of 1, such that $f_{\phi,p}(y^1, y^2, y^3) = \phi(1-p)\phi p$.

Once the probabilities of survival and detection have been estimated, it should, in theory, be possible to calculate life expectancy based on the inferred dead/alive status of individuals, reconstructing the sequence of states for each individual.

This two-state example may be generalized to give a multi-state capture–recapture model (Lebreton *et al.* 2009), incorporating reproductive states. This model may be formulated as a three-state HMM, including a "dead" state and two reproductive states, R and NR: Z^t = R for "alive and reproducing at time t" and Z^t = NR "alive and not reproducing at time t". A schematic representation of this HMM is shown in Figure 3.3.

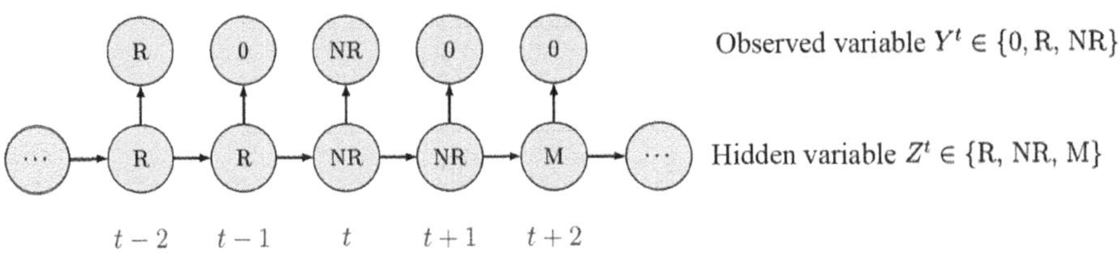

Figure 3.3. *Multi-state capture–recapture model expressed in HMM form*

The first component of the associated HMM is the initial distribution:

$$\boldsymbol{\delta} = \begin{array}{c} \begin{array}{ccc} \text{R} & \text{NR} & \text{M} \end{array} \\ (\delta_R \;\; 1-\delta_R \;\; 0) \end{array}$$

Let ϕ_R be the probability of survival of reproducing individuals, ϕ_{NR} that of non-reproducing individuals, $\psi_{NR,R}$ the probability of an individual, which is not reproducing at time t entering the reproductive state at time $t+1$, and $\psi_{R,NR}$ the probability that an individual, which is in the reproductive state at time t, will have left this state at time $t+1$. The transition matrix is written as:

$$\boldsymbol{\Gamma} = \begin{array}{c} \begin{array}{ccc} \text{R} & \text{NR} & \text{M} \end{array} \\ \begin{bmatrix} \phi_R(1-\psi_{R,NR}) & \phi_R\psi_{R,NR} & 1-\phi_R \\ \phi_{NR}\psi_{NR,R} & \phi_{NR}(1-\psi_{NR,R}) & 1-\phi_{NR} \\ 0 & 0 & 1 \end{bmatrix} \begin{array}{l} \text{R} \\ \text{NR} \\ \text{M} \end{array} \end{array}$$

Finally, let p_R be the probability of detection of reproducing individuals (and p_{NR} that of non-reproducing individuals). The diagonal matrix giving the distribution of observations conditional on states is thus:

$$\mathbf{P}(y^t) = \begin{array}{c} \\ \\ \\ \end{array}\begin{bmatrix} p_R^{I(y^t=R)}(1-p_R)^{I(y^t=0)}0^{I(y^t=NR)} & 0 & 0 \\ 0 & p_{NR}^{I(y^t=NR)}(1-p_{NR})^{I(y^t=0)}0^{I(y^t=R)} & 0 \\ 0 & 0 & I(y^t=0) \end{bmatrix},$$

with columns labelled R, NR, M.

where $I(y^t = k)$ is the indicator function, taking a value of 1 when $y^t = k$ and 0 otherwise. The distribution implied here is a generalization of the Bernoulli distribution for more than two possible outcomes, that is, a categorical (single-trial multinomial) distribution.

For example, to study reproduction costs, ecologists may compare the probability of reproducing in year $t+1$ based on the individual's reproductive, $(\psi_{R,R} = 1 - \psi_{R,NR})$, or non-reproductive, $(\psi_{NR,R})$, state in year t; the differences in survival rates between reproducing (ϕ_R) and non-reproducing (ϕ_{NR}) individuals may also be studied in this way.

While multi-state models were originally developed for use in estimating demographic parameters (survival, movement, etc.), which depend on geographical sites (Brownie *et al.* 1993), there are few real limits to their application in ecology (Gimenez *et al.* 2012).

Once the parameters have been estimated, the subjacent states can be inferred. In this way, it becomes possible to calculate particularly interesting ecological quantities. Examples of this include the sex ratio (Pradel *et al.* 2008), where the states are the sex of individuals; reproductive success over a lifetime (Rouan *et al.* 2009b; Gimenez *et al.* 2012; Desprez *et al.* 2018); or the number of sick individuals (Buzdugan *et al.* 2017) in the case of epidemiological states.

One tacit hypothesis that is inherent in multi-state levels is that the state of an individual can be measured without error. In practice, however, it can be difficult to assign a sure state to individuals, for example when observing reproduction in the field. A reproductive state can be confirmed if a female is seen with one or more young, for example, but if a female is observed alone, status assignment is less certain. The HMM approach takes account of this element of uncertainty in the assignment of states to individuals (Dupuis 1995; Pradel 2005; Gimenez *et al.* 2012), as we shall see in the following example.

3.3.2. *Case study: estimating the prevalence of dog–wolf hybrids with uncertain individual identification*

The points made above can be illustrated using an example, in this case relating to the estimation of the prevalence of hybrids in a wild animal population. Our case study concerns cross-breeding between dogs and wolves in the Tusco-Emilian Apennines National Park, Italy (Santostasi *et al.* 2019). The data were obtained using wolf feces collected from August 2016 and May 2017, from which DNA was extracted, amplified and sequenced (Caniglia *et al.* 2014); using these DNA data, a distinction can be made between wolves, hybrids and animals of uncertain status. There were five capture sessions, each spanning 2 months, featuring samples from 39 individuals (19 wolves, 12 hybrids and eight uncertain). In the original study, the authors compared different models, including or ignoring the difference between hybrid and parental individuals in terms of detection and assignment probabilities.

The possible states included parental $Z^t = P$, hybrid $Z^t = H$ or dead $Z^t = M$, with observations denoted as $y^t = 0$ for undetected, $y^t = 1$ for observed parental, $y^t = 2$ for observed hybrid and $y^t = 3$ for observed, uncertain status. All parameters in the model used here are constant, except for survival, which is state dependent; hence, $\phi_P \neq \phi_H$.

The components used to write the likelihood of the HMM are the initial distribution

$$\boldsymbol{\delta} = \overset{\begin{matrix}\text{P} & \text{H} & \text{M}\end{matrix}}{\begin{pmatrix}\delta_P & 1-\delta_P & 0\end{pmatrix}},$$

the transition matrix

$$\boldsymbol{\Gamma} = \begin{matrix} & \begin{matrix}\text{P} & \text{H} & \text{M}\end{matrix} & \\ & \begin{bmatrix}\phi_P & 0 & 1-\phi_P \\ 0 & \phi_H & 1-\phi_H \\ 0 & 0 & 1\end{bmatrix} & \begin{matrix}\text{P}\\ \text{H}\\ \text{M}\end{matrix}\end{matrix}$$

and the diagonal matrix which gives the distribution of observations conditional to the states

$$\mathbf{P}(y^t) = \begin{matrix} \begin{matrix}\text{P} & \text{H} & \text{M}\end{matrix} \\ \begin{bmatrix} f(y^t|Z^t = P) & 0 & 0 \\ 0 & f(y^t|Z^t = H) & 0 \\ 0 & 0 & I(y^t = 0)\end{bmatrix}\end{matrix},$$

where $f(y^t|Z^t = P) = (1-p)^{I(y^t=0)}(p\delta)^{I(y^t=1)}0^{I(y^t=2)}(p(1-\delta))^{I(y^t=3)}$ and $f(y^t|Z^t = H) = (1-p)^{I(y^t=0)}0^{I(y^t=1)}(p\delta)^{I(y^t=2)}(p(1-\delta))^{I(y^t=3)}$.

The important parameter here is δ, the probability of an individual being assigned to a state. If the genetic or morphological assessment is not sufficient to assign parental or hybrid status to an individual, then it will be classified as uncertain, with probability $1 - \delta$.

As the hybridization test was carried out just once for each genotype, the assignment probability δ is estimated for the first capture alone. The assignment of parental or hybrid status to individuals in the uncertain category, and consequently the calculation of the prevalence of hybrids, is carried out using global decoding by means of the Viterbi algorithm.

The probability of survival for wolves ϕ_P is estimated at 0.63 (0.39–0.82), lower than the probability of survival for hybrids ϕ_H, estimated at 0.81 (0.59–0.93). The probability of detection p is estimated to be 0.46 (0.31–0.61) and the probability of assignment δ is estimated to be 0.85 (0.75–0.91).

The main result in this case is an estimation of the number of hybrid individuals. The estimated prevalence varies from 0.18 to 0.33, and is comparable to the observed prevalence (Table 3.1).

Prevalence	Occ. 1	Occ. 2	Occ. 3	Occ. 4	Occ. 5
Observed	0.27	0.33	0.20	0.46	0.27
Estimated	0.27	0.33	0.20	0.20	0.18
95% Confidence interval	(0.09, 0.61)	(0.10,0.61)	(0.00, 0.50)	(0.00, 0.50)	(0.00, 0.50)

Table 3.1. *Prevalence of hybrids: observed and estimated using the Viterbi algorithm*

Santostasi *et al.* (2019) compare several models; the authors show that the estimated prevalence is systematically lower than the observed prevalence, with important consequences in terms of species management. The HMM permits a confidence interval to be used in conjunction with the estimation of prevalence.

3.4. HMM and species distribution

3.4.1. *General case*

Instead of working at an individual scale, a different perspective can be gained by using detection and non-detection data at species level. This data give us access to spatial information in relation to species and populations, for example, occupancy. In concrete terms, data are obtained by monitoring several spatial units (such as breeding sites or photo traps) where a species may or may not be detected. Occupancy models are used to estimate the proportion of an area occupied by a species, with corrections for imperfect detectability (MacKenzie *et al.* 2018); in dynamic cases, the probabilities of local extinction ϵ and colonization κ are also included. Sites are treated in exactly the same way as individuals using the capture–recapture approach, and occupancy models are thus similar to the capture–recapture models presented in the previous section. Occupancy models can be seen as HMMs (Royle and Kéry 2007; Gimenez *et al.* 2014a) in which the state process governs the dynamics of site states, with $Z^t = \mathrm{O}$ denoting "occupied site" and $Z^t = \mathrm{NO}$ denoting "non-occupied site" for a year t. A species may be detected, $Y^{t,k} = 1$, or undetected, $Y^{t,k} = 0$, at each site on multiple visits k over the course of a year t. A schematic representation of the corresponding HMM is shown in Figure 3.4.

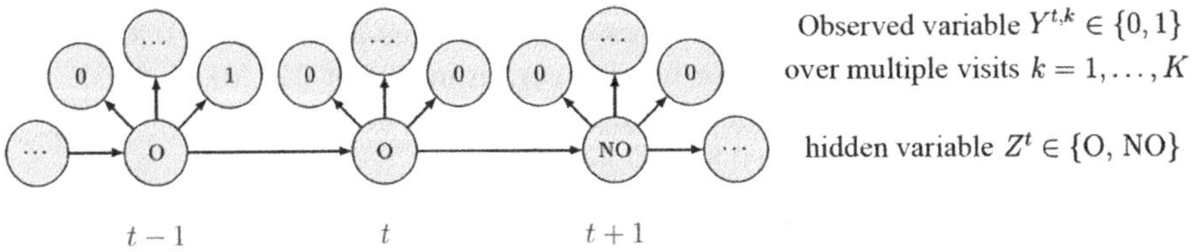

Figure 3.4. *Diagram of a dynamic occupancy model expressed as an HMM*

The components used in constructing the likelihood of the model are written as above. We begin with the initial distribution:

$$\delta = \begin{pmatrix} \overset{\mathrm{O}}{\psi_1} & \overset{\mathrm{NO}}{1-\psi_1} \end{pmatrix},$$

where ψ_1 is the probability of initial occupancy (in the first year). The transition matrix is written as:

$$\boldsymbol{\Gamma} = \begin{array}{c} \begin{matrix} \mathrm{O} & \mathrm{NO} \end{matrix} \\ \begin{bmatrix} 1-\epsilon & \epsilon \\ \kappa & 1-\kappa \end{bmatrix} \end{array} \begin{matrix} \mathrm{O} \\ \mathrm{NO} \end{matrix}$$

Finally, the state-dependent matrix of the observation distribution is:

$$\mathbf{P}(\mathbf{y}^t) = \begin{array}{c} \begin{matrix} \mathrm{O} & \qquad\qquad \mathrm{NO} \end{matrix} \\ \begin{bmatrix} \prod_{k=1}^{K} p^{y^{t,k}} (1-p)^{1-y^{t,k}} & 0 \\ 0 & \prod_{k=1}^{K} (1-y^{t,k}) \end{bmatrix} \end{array},$$

where p is the probability of detection of the species.

One special case is that of single-season (static) occupancy (MacKenzie *et al.* 2002) where $\epsilon = \kappa = 0$ (Gimenez *et al.* 2014a) and $T = 1$. The HMM formulation allows us to estimate not only the probabilities of occupancy, extinction and colonization but also the state of a site if the species has not been detected (via global decoding). Because of the flexibility of the HMM formulation, the standard model can be extended to take account of differences in the probability of detecting a species via finite number mixtures (Louvrier *et al.* 2018a) or a discrete measure of this heterogeneity such as population density or reproductive state (Gimenez *et al.* 2014a; Veran *et al.* 2015); it can also take account of the occurrence of false positives in the data due to erroneous species identification (Miller *et al.* 2011; Louvrier *et al.* 2019). As in the case of multi-state capture–recapture models, HMM occupancy models can be extended to include multiple "occupied" states, such as reproductive states (MacKenzie *et al.* 2009; Martin *et al.* 2009), epidemiological states (McClintock *et al.* 2010) or landscape-related states (Lamy *et al.* 2013). These models can also be extended to cases with multiple species in order to study predator–prey relationships (Rota *et al.* 2016; Fidino *et al.* 2019).

3.4.2. *Case study: estimating the distribution of a wolf population with species identification errors and heterogeneous detection*

In this case, an HMM will be used to model species distribution in a case featuring identification errors and heterogeneous detection. The data analyzed relate to the detection and non-detection of wolves in France, and were collected in 2013 (Louvrier *et al.* 2018b). Signs that the species was present, such as tracks, feces, prey remains, dead animals, camera trap photographs and actual spottings were collected by a network of professional and amateur observers (Duchamp *et al.* 2012). The data for 2013 comprised 250 certain detections, 54 uncertain detections (cases of confusion with another species) and 12,540 non-detections across a grid of 3,211 sites over a 10 x 10 km space. We have considered each month, from December to March, as a separate sampling occasion. These months correspond to a period between two dispersal events, in the fall and the spring (Louvrier *et al.* 2018b). This choice increases the chance of respecting an important hypothesis inherent to occupancy models, namely that the state of the site should stay the same over the course of the study. In a previous study, we found that the main explanatory factor for occupation was site altitude, and that the probability of detecting the species was mostly determined by the sampling effort, defined as the number of observers per site per year (Louvrier *et al.* 2018b). In this case, for illustrative purposes, we have chosen to focus on a model that takes account of identification errors and heterogeneous detection in the determination of detection probabilities (Louvrier *et al.* 2018a). After estimating the parameters of the model, we constructed a map representing the 3,211 sites in the study area, each associated with a heterogeneity class estimated using the Viterbi algorithm.

We considered two classes of sites, A and B, with respective proportions π and $1-\pi$. The possible states were $Z^k = OA$ for an occupied site of class, A, $Z^k = OB$ for an occupied site of class B, $Z^k = NOA$ for a non-occupied site of class A and $Z^k = NOB$ for a non-occupied site of class B. We constructed a single-season (static) model with $k = 1, \ldots, K$ visits. Observations were denoted as $y^k = 0$ for a site where the species was not observed, $y^k = 1$ for a site with an unambiguous observation and $y^k = 2$ for a site with an ambiguous observation. In this case, we use a model in which all parameters are constant over time, but dependent on the site classification in terms of detection.

The components used in writing the likelihood of the HMM are the initial distribution as follows:

$$\begin{array}{lcccc} & \text{NOA} & \text{NOB} & \text{OA} & \text{OB} \\ \boldsymbol{\delta} = & (\pi(1-\psi_A) & (1-\pi)(1-\psi_B) & \pi\psi_A & (1-\pi)\psi_B) \end{array}$$

and the diagonal matrix giving the distribution of observations conditional on states:

$$\mathbf{P}(y^k) = \begin{array}{c} \begin{array}{cccc} \text{NOA} & \text{NOB} & \text{OA} & \text{OB} \end{array} \\ \begin{bmatrix} f(y^k|Z^k = NOA) & 0 & 0 & 0 \\ 0 & f(y^k|Z^k = NOB) & 0 & 0 \\ 0 & 0 & f(y^k|Z^k = OA) & 0 \\ 0 & 0 & 0 & f(y^k|Z^k = OB) \end{bmatrix} \end{array},$$

where $f(y^k|Z^k = NOA) = (1-p_{A10})^{I(y^k=0)}0^{I(y^k=1)}p_{A10}^{I(y^k=2)}$, $f(y^k|Z^k = NOB) = (1-p_{B10})^{I(y^k=0)}0^{I(y^k=1)}p_{B10}^{I(y^k=2)}$, $f(y^k|Z^k = OA) = (1-p_{A11})^{I(y^k=0)}(bp_{A11})^{I(y^k=1)}(1-b)p_{A11})^{I(y^k=2)}$ and $f(y^k|Z^k = OB) = (1-p_{B11})^{I(y^k=0)}(bp_{B11})^{I(y^k=1)}(1-b)p_{B11})^{I(y^k=2)}$ where p_{A11} is the probability of correctly detecting the species at a class A occupied site (respectively, p_{B11} for class B), p_{A10} is the probability of wrongly detecting the species at a class A non-occupied site (respectively, p_{B10} for B), and b is the probability of classifying a true positive as unambiguous or certain. As there is no dynamic element with respect to site state, the transition matrix is the identity matrix.

The model presents several local maxima in the likelihood, something which is common when using HMMs. It can be hard to pinpoint the reason for this problem; our preferred approach is to apply multiple numerical optimizations, changing the initial values each time. In this case, 100 random drawings were carried out from a uniform distribution between 0 and 1 to provide initial values for the model parameters, which are all probabilities; the model was then adjusted for each combination. The results are striking, featuring multiple optima, as shown in Figure 3.5.

The estimated probability of occupation is low, at 0.05 (0.04–0.06). According to the fitted model, 94% have a zero probability of detection of a false positive p_{B10}, indicating that there are no identification errors for these sites. From the remaining 6% of sites, the estimated value of p_{A10} is also low at 0.05 (0.03–0.08). These results suggest that the training procedure followed by observers in the network was effective, and/or that the data filtering process applied prior to analysis minimizes the number of false positives. The probability b of classifying a true positive as non-ambiguous is high, estimated at 0.93 (0.90–0.95). Taken in conjunction with the low risk of false positives, this result suggests that uncertain detections could be considered as certain. Finally, the probability of detection of true positives p_{11} was estimated at 1 for 6% of sites, and at 0.39 (0.35–0.43) for the remaining 94%.

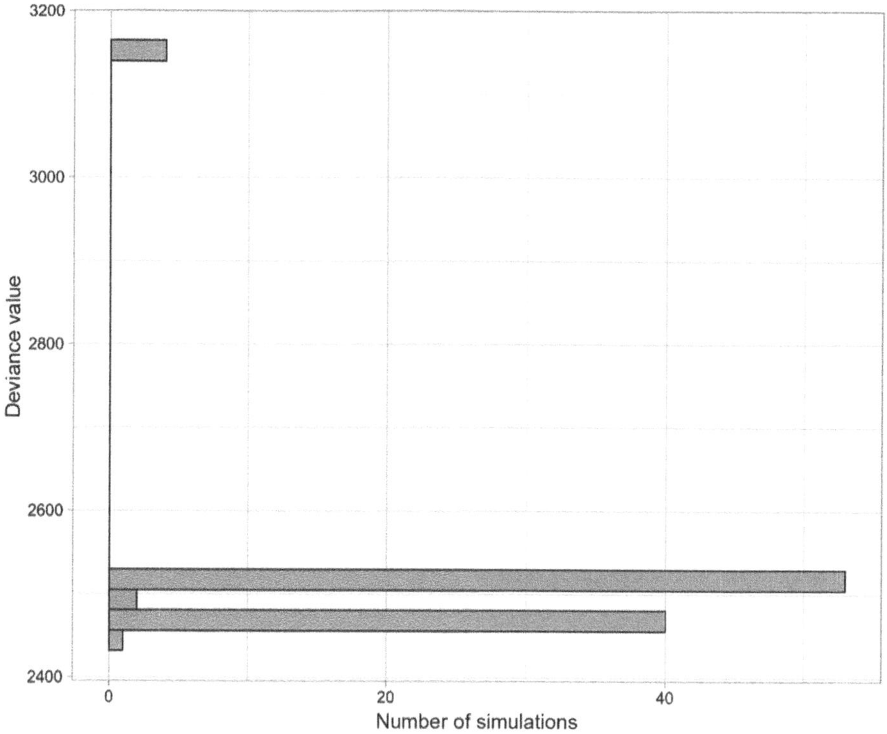

Figure 3.5. *Identification of local minima in the $-2\log(\mathcal{L}(\boldsymbol{\theta}))$ deviance of an HMM. Numerical optimization was carried out using 100 random drawings of initial values. The graph shows number of instances (x axis) against value (y axis). Several local minima are clearly visible*

Once the parameters of the model have been estimated, the Viterbi algorithm may be used to determine the most probable state for each site. Once the most probable state of each site has been determined, the results may be viewed on a map, such as that shown in Figure 3.6, showing the level of heterogeneity. Observed variations between sites are partly the result of spatial variations in sampling effort, defined as the number of active observers for a site (Louvrier *et al.* 2018a). The interest of using HMMs in this case lies in the ability to take account of heterogeneity in the observation process. This is done using a hidden variable to account for belonging to a finite number of classes; it is thus possible to avoid the need to measure sampling effort on the ground, a promising property for analyzing data obtained using participative approaches.

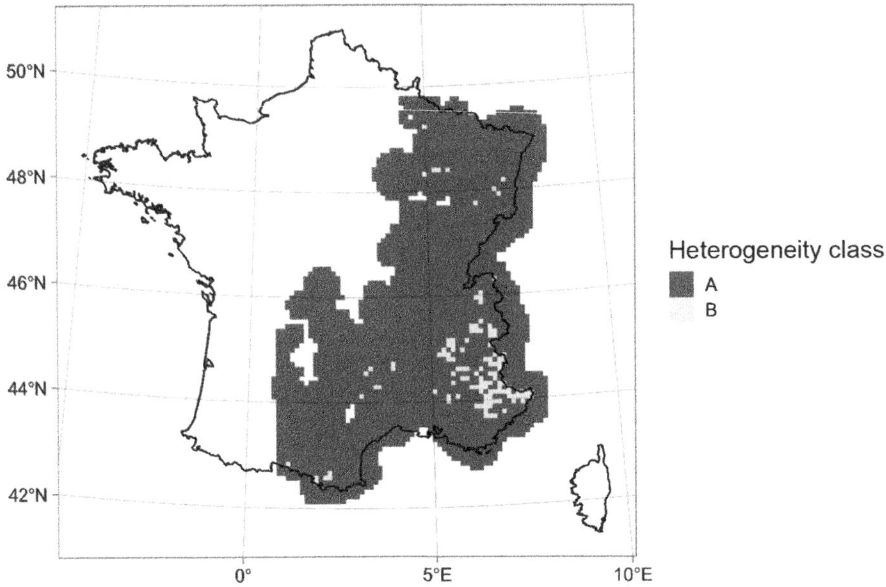

Figure 3.6. *Visualization of heterogeneity: map of the heterogeneity class to which each site in the study area is assigned using the Viterbi algorithm. For a color version of this figure, see www.iste.co.uk/peyrard/ecology.zip*

3.5. Discussion

In this chapter, we have seen how HMMs can be used in ecology to address questions about the demography and distribution of species in their natural environment. The flexibility and ecological relevance of the HMM modeling framework have contributed to its increasing popularity in ecology, where it is used in relation to a wide range of questions (McClintock *et al.* 2020). The main advantage of the HMM approach lies in the ability to infer the ecological states of individuals and species which are, at best, partially observable: these are hidden variables. In addition to the ability to explicitly distinguish between observation processes and states, it is possible to decompose potentially complex processes into several simpler steps (Choquet 2008), facilitating model construction (Louvrier *et al.* 2018a; Santostasi *et al.* 2019). Finally, HMMs make it possible to infer state dynamics in time and space. Note that model selection and approaches to testing the quality of adjustment of models to data are not covered here; for a detailed discussion of these issues, see Zucchini *et al.* (2016) and McClintock *et al.* (2020).

Nevertheless, HMMs do have limitations, three of which will be discussed here. The first limitation is numerical in nature. As we saw in our case study concerning occupancy models, the likelihood function may present local maxima, which makes global maximization complex. The solution to this problem generally involves testing several sets of initial values for numerical optimization, via random drawings, as in the case described above; another option is to use estimated parameters for a simplified model less subject to local maxima as the initial values. Other approaches may also be used (Brooks and Morgan 1994). A further problem is linked to the non-identifiability of models for which the likelihood is uniform in areas, for example in the case of redundant parameters; this problem can be diagnosed (Cole 2019).

The second limitation concerns the Markovian hypothesis itself. This hypothesis implies that the time taken to move from one state to another follows a geometric distribution, and this is not always verified in practice. One solution to this problem is to consider Markov chains with an order greater than 1, or, in other words, to assign memory to HMMs. In terms of demography, this consists, for example, of admitting that the probability of movement between geographical sites depends not only on the current site, but also on previously visited sites (Rouan *et al.* 2009a; Cole *et al.* 2014). Another solution is to model the time spent in a state directly, in the form of a semi-Markov model (Choquet *et al.* 2011; King and Langrock 2016).

The third limitation is regarding the discrete nature of states in HMMs. In cases where a finite number of states are used to approximate the distribution of a continuous variable, such as the mass of an individual or the geographical range of a species, the question of discretization must be addressed. Evidently, the number of states may be increased to make the discretization finer, but at the cost of increased complexity, via an increase in the number of parameters and/or states to estimate. The problems relating to high-dimensional space states can be mitigated by exploiting the fact that only certain transitions are possible, increasing calculation efficiency (Glennie *et al.* 2019); another option is to group states (Besbeas and Morgan 2019).

In this chapter, we have demonstrated the adjustment of HMMs in a frequentist setting, combining an efficient expression of likelihood using the forward algorithm with numerical optimization in order to obtain maximum likelihood estimators of parameters, then using the Viterbi algorithm to reconstruct the most likely sequence of states (hidden variables) in a process known as decoding. Our approach can be implemented in R and is reproducible[1]. There are several computer-based solutions

1 The code is available to download from GitHub: https://oliviergimenez.github.io/code_livre_variables_cachees/gimenez. html.

for implementing a frequentist approach and for using HMMs to analyze capture–recapture or occupancy data (Choquet *et al.* 2009; Fiske and Chandler 2011; Laake 2013; Gimenez *et al.* 2014a). Other tools that may be used in this context include the EM algorithm (see Chapter 5) or the Bayesian approach, implemented via Markov chain Monte Carlo (MCMC) methods. The Bayesian approach is enjoying increasing popularity for adjusting statistical models in the field of ecology, notably due to the availability of flexible, powerful programs (de Valpine *et al.* 2017; Plummer 2003). A major advantage of the Bayesian approach is that hidden variables are treated as parameters to estimate, making it easy to take account of a measure of uncertainty with regard to these variables. However, the drawback is that standard MCMC samplers do not perform particularly well in cases where both parameters and hidden variables must be determined. One solution is to apply sampling to the parameters alone, marginalizing states via the forward algorithm (Turek *et al.* 2016; Yackulic *et al.* 2020), but this has a negative effect on the estimation of hidden variables. Research into the use of the Viterbi algorithm within a Bayesian framework is currently ongoing (Lember *et al.* 2019).

3.6. Acknowledgments

This work was partly funded in the context of the ANR DEMOCOM project (ANR-16-CE02-0007). The authors would like to thank the participants in the "wolf" network under the supervision of the French Office for Biodiversity (Office Français de la Biodiversité); C. Duchamp for assistance and provision of wolf data (France); M. Canestrini and F. Moretti for wolf sample collection (Italy); M. Galaverni for assistance with genetic analysis. Julie Louvrier wishes to thank the Université de Montpellier and the OFB for her thesis grant. Nina Santostasi wishes to thank the University of Rome La Sapienza for her thesis grant.

3.7. References

Amorös, J. (2014). Recapturing laplace. *Significance*, 11(3), 38–39.

Auger-Méthé, M., Newman, K., Cole, D., Empacher, F., Gryba, R., King, A.A., Leos-Barajas, V., Flemming, J.M., Nielsen, A., Petris, G., Thomas, L. (2020). An introduction to state-space modeling of ecological time series. arXiv preprint arXiv:2002.02001.

Besbeas, P. and Morgan, B.J. (2019). Exact inference for integrated population modelling. *Biometrics*, 75(2), 475–484.

Brooks, S.P. and Morgan, B.J. (1994). Automatic starting point selection for function optimization. *Statistics and Computing*, 4(3), 173–177.

Brownie, C., Hines, J.E., Nichols, J.D., Pollock, K.H., Hestbeck, J.B. (1993). Capture–recapture studies for multiple strata including non-Markovian transitions. *Biometrics*, 49, 1173–1187.

Buzdugan, S., Vergne, T., Grosbois, V., Delahay, G., Drewe, J. (2017). Inference of the infection status of individuals using longitudinal testing data from cryptic populations: Towards a probabilistic approach to diagnosis. *Scientific Reports*, 7(1111) [Online]. Available at: https://www.nature.com/articles/s41598-017-00806-4.

Caniglia, R., Fabbri, E., Galaverni, M., Milanesi, P., Randi, E. (2014). Noninvasive sampling and genetic variability, pack structure, and dynamics in an expanding wolf population. *Journal of Mammalogy*, 95(1), 41–59.

Choquet, R. (2008). Automatic generation of multistate capture–recapture models. *The Canadian Journal of Statistics/La Revue Canadienne de Statistique*, 36(1), 43–57.

Choquet, R., Rouan, L., Pradel, R. (2009). Program E-SURGE: A software application for fitting multievent models. In *Modeling Demographic Processes in Marked Populations*, Thomson, D.L., Cooch, E.G., Conroy, M.J. (eds). Springer, Boston, MA.

Choquet, R., Viallefont, A., Rouan, L., Gaanoun, K., Gaillard, J.-M. (2011). A semi-Markov model to assess reliably survival patterns from birth to death in free-ranging populations. *Methods in Ecology and Evolution*, 2(4), 383–389.

Choquet, R., Carrié, C., Chambert, T., Boulinier, T. (2013). Estimating transitions between states using measurements with imperfect detection: Application to serological data. *Ecology*, 94(10), 2160–2165.

Choquet, R., Garnier, A., Awuve, E., Besnard, A. (2017). Transient state estimation using continuous-time processes applied to opportunistic capture–recapture data. *Ecological Modelling*, 361, 157–163.

Cole, D.J. (2019). Parameter redundancy and identifiability in hidden Markov models. *METRON*, 77(2), 105–118.

Cole, D.J., Morgan, B.J.T., McCrea, R.S., Pradel, R., Gimenez, O., Choquet, R. (2014). Does your species have memory? Analyzing capture–recapture data with memory models. *Ecology and Evolution*, 4(11), 2124–2133.

Desprez, M., Gimenez, O., McMahon, C.R., Hindell, M.A., Harcourt, R.G. (2018). Optimizing lifetime reproductive output: Intermittent breeding as a tactic for females in a long-lived, multiparous mammal. *Journal of Animal Ecology*, 87(1), 199–211.

Duchamp, C., Boyer, J., Briaudet, P.-E., Leonard, Y., Moris, P., Bataille, A., Dahier, T., Delacour, G., Millisher, G., Miquel, C., Poillot, C., Marboutin, E. (2012). Wolf monitoring in France: A dual frame process to survey time- and space-related changes in the population. *Hystrix, the Italian Journal of Mammalogy*, 23(1), 14–28.

Dupont, P., Pradel, R., Lardy, S., Allainé, D., Cohas, A. (2015). Litter sex composition influences dominance status of alpine marmots (*Marmota marmota*). *Oecologia*, 179(3), 753–763.

Dupuis, J. (1995). Bayesian estimation of movement and survival probabilities from capture–recapture data. *Biometrika*, 82(4), 761–772.

Fidino, M., Simonis, J.L., Magle, S.B. (2019). A multistate dynamic occupancy model to estimate local colonization-extinction rates and patterns of co-occurrence between two or more interacting species. *Methods in Ecology and Evolution*, 10(2), 233–244.

Fiske, I. and Chandler, R. (2011). Unmarked: An R package for fitting hierarchical models of wildlife occurrence and abundance. *Journal of Statistical Software*, 43(10), 1–23.

Gimenez, O., Rossi, V., Choquet, R., Dehais, C., Doris, B., Varella, H., Vila, J.-P., Pradel, R. (2007). State-space modelling of data on marked individuals. *Ecological Modelling*, 206(3), 431–438.

Gimenez, O., Lebreton, J.-D., Gaillard, J.-M., Choquet, R., Pradel, R. (2012). Estimating demographic parameters using hidden process dynamic models. *Theoretical Population Biology*, 82(4), 307–316.

Gimenez, O., Blanc, L., Besnard, A., Pradel, R., Doherty Jr, P.F., Marboutin, E., Choquet, R. (2014a). Fitting occupancy models with e-surge: Hidden Markov modelling of presence–absence data. *Methods in Ecology and Evolution*, 5(6), 592–597.

Gimenez, O., Buckland, S.T., Morgan, B.J.T., Bez, N., Bertrand, S., Choquet, R., Dray, S., Etienne, M.-P., Fewster, R., Gosselin, F., Mérigot, B., Monestiez, P., Morales, J.M., Mortier, F., Munoz, F., Ovaskainen, O., Pavoine, S., Pradel, R., Schurr, F.M., Thomas, L., Thuiller, W., Trenkel, V., de Valpine, P., Rexstad, E. (2014b). Statistical ecology comes of age. *Biology Letters*, 10(4) [Online]. Available at: https://royalsocietypublishing.org/doi/full/10.1098/rsbl.2014.0698.

Glennie, R., Borchers, D.L., Murchie, M., Harmsen, B., Foster, R. (2019). Open population maximum likelihood spatial capture–recapture. *Biometrics*, 75, 1345–1355.

Grinnell, J. (1917). The niche-relationships of the California thrasher. *The Auk*, 34(4), 427–433.

Johnson, D.S., Laake, J.L., Melin, S.R., DeLong, R.L. (2016). Multivariate state hidden Markov models for mark-recapture data. *Statistical Science*, 31(2), 233–244.

King, R. and Langrock, R. (2016). Semi-Markov Arnason–Schwarz models. *Biometrics*, 72(2), 619–628.

Laake, J.L. (2013). Capture–recapture analysis with hidden Markov models. AFSC Processed Report 2013-04, Alaska Fisheries Science Center, NOAA, National Marine Fisheries Service, Seattle, WA.

Laake, J.L., Johnson, D., Diefenbach, D., Ternent, M. (2014). Hidden Markov model for dependent mark loss and survival estimation. *Journal of Agricultural, Biological, and Environmental Statistics*, 19(1), 522–538.

Lamy, T., Gimenez, O., Pointier, J.-P., Jarne, P., David, P. (2013). Metapopulation dynamics of species with cryptic life stages. *American Naturalist*, 181(4), 479–491.

Lebreton, J.-D., Nichols, J.D., Barker, R.J., Pradel, R., Spendelow, J.A. (2009). Modeling individual animal histories with multistate capture–recapture models. *Advances in Ecological Research*, 41, 87–173.

Lember, J., Gasbarra, D., Koloydenko, A., Kuljus, K. (2019). Estimation of Viterbi path in Bayesian hidden Markov models. *METRON*, 77, 137–169.

Louvrier, J., Chambert, T., Marboutin, E., Gimenez, O. (2018a). Accounting for misidentification and heterogeneity in occupancy studies using hidden Markov models. *Ecological Modelling*, 387, 61–69.

Louvrier, J., Duchamp, C., Lauret, V., Marboutin, E., Cubaynes, S., Choquet, R., Miquel, C., Gimenez, O. (2018b). Mapping and explaining wolf recolonization in France using dynamic occupancy models and opportunistic data. *Ecography*, 41(4), 647–660.

Louvrier, J., Molinari-Jobin, A., Kéry, M., Chambert, T., Miller, D., Zimmermann, F., Marboutin, E., Molinari, P., Meller, O., Arne, R., Gimenez, O. (2019). Use of ambiguous detections to improve estimates from species distribution models. *Conservation Biology: The Journal of the Society for Conservation Biology*, 33(1), 185–195.

MacKenzie, D.I., Nichols, J.D., Lachman, G.B., Droege, S., Andrew Royle, J., Langtimm, C.A. (2002). Estimating site occupancy rates when detection probabilities are less than one. *Ecology*, 83(8), 2248–2255.

MacKenzie, D.I., Nichols, J.D., Seamans, M.E., Gutiérrez, R. (2009). Modeling species occurrence dynamics with multiple states and imperfect detection. *Ecology*, 90(3), 823–835.

MacKenzie, D.I., Nichols, J.D., Royle, J.A., Pollock, K.H., Bailey, L., Hines, J.E. (2018). *Occupancy Estimation and Modeling: Inferring Patterns and Dynamics of Species Occurrence*, 2nd edition Elsevier, Cambridge, MA.

Marescot, L., Benhaiem, S., Gimenez, O., Hofer, H., Lebreton, J.-D., Olarte-Castillo, X.A., Kramer-Schadt, S., East, M.L. (2018). Social status mediates the fitness costs of infection with canine distemper virus in Serengeti spotted hyenas. *Functional Ecology*, 32(5), 1237–1250.

Martin, J., McIntyre, C.L., Hines, J.E., Nichols, J.D., Schmutz, J.A., MacCluskie, M.C. (2009). Dynamic multistate site occupancy models to evaluate hypotheses relevant to conservation of golden eagles in Denali National Park, Alaska. *Biological Conservation*, 142(11), 2726–2731.

McClintock, B.T., Nichols, J.D., Bailey, L.L., MacKenzie, D.I., Kendall, W.L., Franklin, A.B. (2010). Seeking a second opinion: Uncertainty in disease ecology. *Ecology Letters*, 13(6), 659–674.

McClintock, B.T., Langrock, R., Gimenez, O., Cam, E., Borchers, D.L., Glennie, R., Patterson, T.A. (2020). Uncovering ecological state dynamics with hidden Markov models. *Ecology Letters*, 23(12), 1878–1903.

Mews, S., Langrock, R., King, R., Quick, N. (2020). Continuous-time multi-state capture–recapture models. arXiv preprint arXiv:2002.10997.

Miller, D.A., Nichols, J.D., McClintock, B.T., Grant, E.H.C., Bailey, L.L., Weir, L.A. (2011). Improving occupancy estimation when two types of observational error occur: Non-detection and species misidentification. *Ecology*, 92(7), 1422–1428.

Nichols, J.D., Hines, J.E., Pollock, K.H., Hinz, R.L., Link, W.A. (1994). Estimating breeding proportions and testing hypotheses about costs of reproduction with capture–recapture data. *Ecology*, 75(7), 2052–2065.

Plummer, M. (2003). JAGS: A program for analysis of Bayesian graphical models using Gibbs sampling. *Proceedings of the 3rd International Workshop on Distributed Statistical Computing*, Vienna, March 20–22.

Pradel, R. (2005). Multievent: An extension of multistate capture–recapture models to uncertain states. *Biometrics*, 61, 442–447.

Pradel, R., Maurin-Bernier, L., Gimenez, O., Genovart, M., Choquet, R., Oro, D. (2008). Estimation of sex-specific survival with uncertainty in sex assessment. *Canadian Journal of Statistics*, 36(1), 29–42.

Rabiner, L.R. (1989). A tutorial on hidden Markov models and selected applications in speech recognition. *Proceedings of the IEEE*, 77(2), 257–286.

Rota, C.T., Ferreira, M.A.R., Kays, R.W., Forrester, T.D., Kalies, E.L., McShea, W.J., Parsons, A.W., Millspaugh, J.J. (2016). A multispecies occupancy model for two or more interacting species. *Methods in Ecology and Evolution*, 7(10), 1164–1173.

Rouan, L., Choquet, R., Pradel, R. (2009a). A general framework for modeling memory in capture–recapture data. *Journal of Agricultural, Biological, and Environmental Statistics*, 14, 338–355.

Rouan, L., Gaillard, J.-M., Guédon, Y., Pradel, R. (2009b). Estimation of lifetime reproductive success when reproductive status cannot always be assessed. In *Modeling Demographic Processes in Marked Populations, Environmental and Ecological Statistics*, Thomson, D.L., Cooch, E.G., Conroy, M.J. (eds). Springer, Boston, MA.

Royle, J.A. and Dorazio, R.M. (2008). *Hierarchical Modeling and Inference in Ecology: The Analysis of Data from Populations, Metapopulations and Communities*. Elsevier, Cambridge, MA.

Royle, J.A. and Kéry, M. (2007). A Bayesian state-space formulation of dynamic occupancy models. *Ecology*, 88(7), 1813–23.

Santostasi, N.L., Ciucci, P., Caniglia, R., Fabbri, E., Molinari, L., Reggioni, W., Gimenez, O. (2019). Use of hidden Markov capture–recapture models to estimate abundance in the presence of uncertainty: Application to the estimation of prevalence of hybrids in animal populations. *Ecology and Evolution*, 9(2), 744–755.

Turek, D., de Valpine, P., Paciorek, C.J. (2016). Efficient Markov chain Monte Carlo sampling for hierarchical hidden Markov models. *Environmental and Ecological Statistics*, 23(4), 549–564.

de Valpine, P., Turek, D., Paciorek, C.J., Anderson-Bergman, C., Lang, D.T., Bodik, R. (2017). Programming with models: Writing statistical algorithms for general model structures with NIMBLE. *Journal of Computational and Graphical Statistics*, 26(2), 403–413.

Veran, S., Simpson, S.J., Sword, G.A., Deveson, E., Piry, S., Hines, J.E., Berthier, K. (2015). Modeling spatiotemporal dynamics of outbreaking species: Influence of environment and migration in a locust. *Ecology*, 96(3), 737–748.

Yackulic, C.B., Dodrill, M., Dzul, M., Sanderlin, J.S., Reid, J.A. (2020). A need for speed in Bayesian population models: A practical guide to marginalizing and recovering discrete latent states. *Ecological Applications*, 30, e02112.

Zucchini, W., MacDonald, I.L., Langrock, R. (2016). *Hidden Markov Models for Time Series: An Introduction Using R*. CRC Press, Boca Raton, FL.

4

Inferring Mechanistic Models in Spatial Ecology Using a Mechanistic-Statistical Approach

Julien PAPAÏX[1]**, Samuel SOUBEYRAND**[1]**, Olivier BONNEFON**[1]**, Emily WALKER**[1]**, Julie LOUVRIER**[2]**, Etienne KLEIN**[1] **and Lionel ROQUES**[1]

[1] *Biostatistique et Processus Spatiaux (BioSP), INRAE, Avignon, France*

[2] *Department of Ecological Dynamics, Leibniz Institute for Zoo and Wildlife Research, Berlin, Germany*

4.1. Introduction

Understanding where and when a species is present is crucial in ecology (Elith and Leathwick 2009). The models used are generally based on correlations between data representing the spatio-temporal dynamics of an organism and covariates representing the biotic and abiotic environment in which it exists (Guisan and Thuiller 2005). Outputs of such models are often interpreted as ecological processes that explain the species distributions. However, these correlations do not necessarily indicate a causality relationship (Hefley *et al.* 2017). To overcome this issue, specific mechanisms can be considered explicitly in population dynamics models.

Within the context of hierarchical modeling, such mechanisms are integrated in the modeling of latent variables that describe the dynamics of the studied population. A variety of different deterministic and stochastic models have been developed. In

Statistical Models for Hidden Variables in Ecology,
coordinated by Nathalie PEYRARD and Olivier GIMENEZ.

this chapter, we focus on the use of deterministic models (for more details on the subject of stochastic mechanistic models, see Soubeyrand *et al.* (2009) or Papaïx *et al.* (2021)) based on differential or partial differential equations (PDEs), with a special focus on reaction–diffusion models. These models have become increasingly widespread in ecology over the past decades, notably thanks to the work of Shigesada and Kawasaki (1997), Turchin (1998) and Murray (2002). They provide a means of representing both growth and dispersal processes in heterogeneous environments, while remaining parsimonious. The parsimonious nature of the models makes them easy to simulate and estimate, a considerable advantage in the context of hierarchical modeling. In the following sections, we present a rigorous means of defining these models mathematically and show how they can be solved numerically. Parameters inference rely on the definition of a probabilistic observation model depending on the outputs of the mechanistic model. We present an approach to establishing the link between latent variables and data, which are generally noisy, partial and/or non-commensurable. A description of suitable estimation methods that may be used in this context is also given. Finally, the use of this approach is illustrated through three case studies.

4.2. Dynamic systems in ecology

4.2.1. *Temporal models*

Ordinary differential equations (ODEs) are used in population dynamics to describe the evolution of the size of one or more populations of individuals over time. In these models, time is the only variable. In each of the modeled populations, each individual interacts with all other individuals.

One of the main strengths of these models lies in the ease of implementation and numerical solution, even in cases with a large number of interacting populations.

Single-population models: The simplest ODE models are described by a single equation of the form

$$\begin{cases} N'(t) = f(N(t)), \; t \in [0, T[, \\ N(0) = N_0 \geq 0, \end{cases} \qquad [4.1]$$

where the unknown variable is the size of the population $N(t)$ at each instant $t > 0$. Quantity N_0 corresponds to the initial size of the population. The function $f \in C^1(\mathbb{R})$ is the *growth function* of the population. The instantaneous variation in population size is given by

$$N'(t) = \text{nb births } t^{-1} - \text{nb deaths } t^{-1} \qquad [4.2]$$

Supposing that the numbers of births and deaths are proportional to the size of the population, we obtain:

$$N'(t) = b\,N(t) - d\,N(t) = (b-d)N(t)\,,$$

where $b > 0$ and $d > 0$ correspond to birth and death rates, respectively, and where $r = b - d$ is the population growth rate. Taking N_0 as the initial population (i.e. at $t = 0$), we obtain:

$$N(t) = N_0 e^{r\,t}\,.$$

This is the simplest ODE model, corresponding to the Malthusian growth model. It may be used, for example, to model the early stages of colonization when the number of individuals is still low. However, it is generally unrealistic over longer periods when $r > 0$: it fails to account for inter-individual competition for resources and predicts exponential population growth.

Equation [4.2] can be modified to account for competition between individuals. In this case, let us suppose that the death rate increases with population size: we replace d by $d(N) = d_0 + d_1\,N$, where $d_1 > 0$. Thus, we obtain:

$$N'(t) = bN - (d_0 + d_1 N)N\,,$$

i.e.

$$N'(t) = (b - d_0)N\left[1 - \frac{d_1}{b - d_0}N\right],$$

Taking $r = b - d_0 > 0$ and $K = \dfrac{b - d_0}{d_1}$, we obtain:

$$N'(t) = rN\left(1 - \frac{N}{K}\right),\ t \geq 0, \qquad [4.3]$$

which is the logistic equation in its classic form. Coefficient K is the carrying capacity of the habitat (expressed as a number of individuals). The solution to model [4.3] can be calculated analytically:

$$N(t) = \frac{KN_0 e^{rt}}{K + N_0(e^{rt} - 1)},\ \text{for } t \geq 0. \qquad [4.4]$$

Note that, irrespective of the initial value of N_0, the population $N(t)$ converges to K over a sufficiently long period of time.

In a model $N' = f(N)$, the function $f(N)/N$ is known as the *per capita* growth rate. This function corresponds to the mean growth rate per individual. In the logistic case, we have $f(N)/N = r(1 - N/K)$. Thus, the *per capita* growth rate increases as the population size decreases. This approach may not be entirely realistic in certain situations, for example, where there is a form of cooperation between individuals. The term *Allee effect*, named in homage to Allee's seminal work on inter-individual cooperation (Allee 1931), is used to refer to cases where $f(N)/N$ does not reach its maximum value when $N \to 0$. This frequently occurs in sexual populations, for example, due to the difficulty of finding a partner in cases of low population density. A negative $f(N)/N$ for small values of N reflects a strong Allee effect, while in cases where $f(N)/N$ does not reach its maximum value at 0 but remains positive, we speak of a weak Allee effect. The growth function

$$f(N) = rN\left(1 - \frac{N}{K}\right)\left(N - \rho\right) \qquad [4.5]$$

is the archetypal growth function inducing an Allee effect, although other examples do exist. In this case, if $\rho > 0$, the Allee effect is strong; if $\rho \in (-K/2, 0]$, the Allee effect is weak, and if $\rho \leq -K/2$, there is no Allee effect. Contrary to what we observed in equations [4.1] and [4.3] in the case of a logistic growth term, the behavior of $N(t)$ in this case is strongly dependent on N_0: a population with an initial size strictly lower than ρ will converge to 0. On the other hand, if the initial population size is strictly greater than ρ, it will converge to K, as in the logistic case.

Stochastic models: in birth and death processes, the population N is incremented by $+1$ for a birth event and -1 for a death event. With a birth rate $b\,N$ and a death rate $d\,N$, as the initial population N_0 tends to infinity, the process $N(t)/N_0$ will converge in distribution toward the solution of $X'(t) = (b - d)\,X(t)$, with $X(0) = 1$, due to the law of large numbers. The Malthus equation $N'(t) = r\,N(t)$ thus emerges as the limit of a birth/death process. Now, let us suppose that the birth and death rates are $b = k + \frac{b_1}{N}$ and $d = k + \frac{d_1}{N}$, $b_1, d_1 \geq 0$ and that $k > 0$. In this case, the law of large numbers implies that the process $N(t)/N_0$ converges in distribution toward 1. Over a longer period of time, however, we see that $N(N_0\,t)/N_0$ converges toward the solution of a stochastic differential equation (SDE) (Baake and Wakolbinger 2015; Feller 1951):

$$dN(t) = r\,N(t)dt + \sqrt{\sigma^2 N(t)}dW(t),\ N(0) = N_0 \geq 0 \qquad [4.6]$$

with $r = b_1 - d_1$ and $\sigma^2 = 2k$ the reproductive variance. The term $W(t)$ corresponds to a standard Brownian motion. This type of equation is particularly suitable for describing the dynamics of small populations, and can also be used to describe extinction events, calculate probabilities of extinction, etc. More generally, the deterministic growth term $r\,N(t)$ may be replaced by the growth terms $f(N)$ described above to take account of inter-individual interactions.

Systems with multiple interacting populations: These models are also of the form

$$\begin{cases} N'(t) = F(N(t)),\ t \in [0, T[, \\ N(0) = N_0. \end{cases} \qquad [4.7]$$

In this case, the unknown $N = (N_1, \ldots, N_k)$ is a vector in $\mathbb{R}^k$ $(k \geq 2)$ describing the sizes of k populations, which may interact. The function $F = (f_1, \ldots, f_k) \in C^1(\mathbb{R}^k, \mathbb{R}^k)$ describes the growth of each population, along with the interactions between these populations. Different types of interactions between populations can be described by acting on the form of the function F.

Predator–prey type models describe the interactions between a prey and a predator population. The unknown values are the number of individuals $N_1(t)$ in the prey population and the number of individuals $N_2(t)$ in the predator population. The following model was initially described by Lotka and Volterra:

$$\begin{cases} N_1' = r_1\,N_1 - \alpha_{12}\,N_1\,N_2, \\ N_2' = -r_2\,N_2 + \alpha_{21}\,N_1\,N_2 \end{cases} \qquad [4.8]$$

with r_1, $r_2 > 0$ and α_{12}, $\alpha_{21} > 0$. In this case, in the absence of predators, the prey population exhibits Malthusian growth. In the absence of prey, the predator population decreases exponentially. Interactions are described using the notion that predation reduces the prey growth rate in proportion to the number of predators N_2, and that predation increases the predator growth rate in proportion to the number of prey N_1. This model, while simplistic, is theoretically valuable; it notably shows that periodic dynamics can be obtained without necessarily including time-periodic coefficients.

In Lotka–Volterra type competition models, the unknown values $N_i(t)$, for $i = 1, \ldots, k$, correspond to competing populations. As in the case of predator–prey models, each equation includes both an intraspecific interaction term and an interspecific interaction term:

$$N_i' = r_i N_i \left(1 - \sum_{j=1}^{k} \alpha_{ij} N_j\right) \qquad [4.9]$$

with $r_i > 0$ (intrinsic growth rate of the population i) and $\alpha_{ij} \geq 0$ (effect of competition from population j on population i). The difference between competition and predator–prey models thus lies in the sign of the interaction term $-\alpha_{ij} N_i N_j$, which is always negative.

SIR models are a form of ODE model widely used in epidemiology. SIR models are *compartmental*, dividing the population into multiple classes. The classic example features the classes susceptible, infectious or removed resistant (immune, or deceased: recovered, or removed). The simplest example of the SIR model is as follows:

$$\begin{cases} S'(t) = -\dfrac{\alpha}{N} S(t) I(t), \\ I'(t) = \dfrac{\alpha}{N} S(t) I(t) - \beta I(t), \\ R'(t) = \beta I(t) \end{cases} \qquad [4.10]$$

with $N = S + I + R$ the size of the population (constant). In this case, susceptible individuals (S) are infected at a rate αI proportional to the number of infected individuals (I). Infected individuals become immune at a constant rate β (where $1/\beta$ corresponds to the average period of contagion).

4.2.2. *Spatio-temporal models without reproduction*

In addition to time t, Spatio-temporal models feature a space variable $\mathbf{x} \in \mathbb{R}^d$. The modeled quantity is no longer the size of the population $N(t)$, but rather the density of the population $u(t, \mathbf{x})$ (number of individuals per unit of area in dimension 2).

Diffusion-transport equation: working within $\mathbb{R}^2$, given the initial position of an individual, we wish to calculate the probability $p(t, x_1, x_2)$ of finding this individual at a position $(x_1, x_2) \in \mathbb{R}^2$ at time $t > 0$. Let us begin by supposing that movements are discrete in time with a step τ. During a time step $dt \ll 1$, we consider that the probability of an individual migrating is $M\,dt$. If it migrates, the probability of finding it at a position (x_1+y_1, x_2+y_2) is given by a dispersal kernel $J(y_1, y_2)$. We obtain the following recursion relation: the probability that an individual is at a position (x_1, x_2) at time t is equal to the probability that it was at (x_1, x_2) at time t and did not move, plus all possible contributions from starting points $(x_1 - y_1, x_2 - y_2)$ in $\mathbb{R}^2$:

$$\begin{aligned} p(t + dt, x_1, x_2) &= (1 - M\,dt)\, p(t, x_1, x_2) \\ &\quad + M\,dt \int_{\mathbb{R}^2} J(y_1, y_2)\, p(t, x_1 - y_1, x_2 - y_2)\, dy_1 dy_2, \end{aligned}$$

that is, with $dt \to 0$,

$$\frac{\partial p}{\partial t}(t, x_1, x_2) = M\,(J \star p - p)(t, x_1, x_2), \qquad [4.11]$$

where $\partial p/\partial s$ denotes the partial derivative of the function p with respect to the variable s and $\star$ is the standard convolution product in $\mathbb{R}^2$. Model [4.11] is an *integro-differential* model, and is particularly suited to the study of long-distance dispersion phenomena, using kernels J which decrease slowly when $\|x\| \to \infty$. However, it is difficult to simulate.

Now, let us suppose that the dispersal kernel corresponds to a multivariate normal distribution $\mathcal{N}((\mu_1, \mu_2), \Lambda)$, with mean (μ_1, μ_2) and diagonal covariance matrix $\Lambda = \text{diag}\,(\lambda, \lambda)$.

Applying a Taylor series development when $\lambda \ll 1$, we obtain:

$$\frac{\partial p}{\partial t}(t, x_1, x_2) \approx -\mathbf{v} \cdot \nabla p + D\,\Delta p \qquad [4.12]$$

with

$$\nabla p := \left(\frac{\partial p}{\partial x_1}, \frac{\partial p}{\partial x_2}\right) \text{ and } \Delta p = \frac{\partial^2 p}{\partial x_1^2} + \frac{\partial^2 p}{\partial x_2^2}.$$

The term Δp is the Laplace operator. The coefficient $D = M\,\lambda/2$ is the diffusion coefficient (expressed in (unit of space)^2/(unit of time)), and the vector $\mathbf{v} = (M\mu_1, M\mu_2)$ is the transport coefficient (unit of space)/(unit of time).

In equation [4.12], the quantity $p(t, x_1, x_2)$ is interpreted as the probability density corresponding to the random variable representing the position of the individual at time t. Considering a population of size N_0 made up of independent individuals, distributed according to a density probability p_0 at time $t = 0$, when $N_0 \to \infty$, the distribution of individuals converges toward the solution $u(t, \mathbf{x})$ of the following equation:

$$\begin{cases} \frac{\partial u}{\partial t}(t, \mathbf{x}) = -\mathbf{v} \cdot \nabla u + D\Delta u(t, \mathbf{x}),\ t > 0,\ \mathbf{x} \in \mathbb{R}^2, \\ u_0(\mathbf{x}) := u(0, \mathbf{x}) = N_0\, p_0(\mathbf{x}) \end{cases} \qquad [4.13]$$

This equation can be generalized, for example, to take account of spatial and temporal heterogeneity:

$$\frac{\partial u}{\partial t}(t, \mathbf{x}) = -\mathbf{1} \cdot \nabla(M\,\mu_1 u, M\,\mu_2 u) + \Delta\left(D(t, x_1, x_2)\, u\right),\ t > 0,\ (x, y) \in \mathbb{R}^2 [4.14]$$

with $\mathbf{1} = (1, 1)$ and $D(t, x_1, x_2) = M(t, x_1, x_2)\,\lambda(t, x_1, x_2)/2$. Note that the coefficients are included within the differential operators, instead of taking the form of factors. *Bounded domains*: In practice, equation [4.13] is generally used within a bounded subdomain $\Omega \subset \mathbb{R}^d$. In this case, the behavior of the solution on the boundary $\partial\Omega$ of the domain must be described in order for the problem to be correctly posed, as there are as many solutions as there are boundary conditions.

Two conditions are generally used in population dynamics. In the case of the absorbing condition (Dirichlet), we take $u(t, \mathbf{x}) = 0$ for all $t > 0$ and all $\mathbf{x} \in \partial\Omega$. This implies that the environment outside of the region Ω is so hostile that any organism moving outside of the boundary will die instantly. In the case of the reflecting condition, the flow of organisms across the boundary is considered to be null, either because the barrier is impenetrable and no individual can leave the domain, or because equal numbers of individuals move in and out. To guarantee that the total population size remains the same (conservativeness of the system), we assume that

$$D(t, \mathbf{x})\nabla u \cdot \nu(t, \mathbf{x}) - \mathbf{v}(t, \mathbf{x}) \cdot \nu(\mathbf{x})\, u(t, \mathbf{x}) = 0,\ t > 0,\ \mathbf{x} \in \partial\Omega, \qquad [4.15]$$

where $\nu(\mathbf{x})$ is the outward normal vector to $\partial\Omega$ at point $\mathbf{x}$. In the specific case where there is no transport ($\mathbf{v} \equiv 0$) and diffusion is homogeneous ($D(t, \mathbf{x}) \equiv D_0$), this condition is known as the Neumann condition, and is written as

$$\frac{\partial u}{\partial \nu}(t, \mathbf{x}) := \nabla u(t, \mathbf{x}) \cdot \nu(\mathbf{x}) = 0 \text{ for all } t > 0 \text{ and all } \mathbf{x} \in \partial\Omega.$$

4.2.3. *Spatio-temporal models with reproduction*

Diffusion-transport equations conserve mass, whether in $\mathbb{R}^d$ or in a bounded domain with a reflecting boundary condition. In other terms, the equation describes the dynamics of the density of the population as a result of individual movement, but the size of the population itself, N, remains constant. Inversely, in ODE models of the form $N'(t) = f(N)$, the size of the population is not constant unless $N(0)$ is a zero of the function f. Reaction–diffusion models combine these two approaches, describing both individual movement and population growth. They take the form:

$$\frac{\partial u}{\partial t}(t, \mathbf{x}) = -\mathbf{1} \cdot \nabla(\mathbf{v}(t, \mathbf{x})\, u) + \Delta\,(D(t, \mathbf{x})\, u) + f(t, \mathbf{x}, u) \qquad [4.16]$$

and may be used in bounded domains (with the boundary conditions described above) or in $\mathbb{R}^d$. The growth (or reaction) terms $f(t, \mathbf{x}, u)$ take the same global form as those described above in the context of ODEs. Nevertheless, in this case they may depend on a position in space, as the environment may be more or less favorable to reproduction. Once again, equations may be combined to describe interactions between populations (competition, predation, SIR, etc.).

4.2.4. *Numerical solution*

The reaction–diffusion model [4.16] is simulated by producing a numerical approximation of the solution of the PDE. The first step is to choose a suitable numerical method. These fall into two broad types: interpolation methods, where an approximation is calculated over the whole space as a combination of base functions, and methods which approximate the solution at discretization points. The numerical method relies on a spatial discretization of the zone of study. The geometry of this space may be more or less complex, ranging from a simple geometric form (square, circle, etc.) to the outlines of a geographical or administrative entity (country, region, etc.). Implementing a discretization algorithm over these zones is technically complex. As this step also affects the final simulation, it is crucial to ensure that it is precise enough to approximate the solution to the model. One method of verification is to ensure that the granularity of the discretization does not result in a significant change in the simulation. The choice of time-step algorithm and the resolution of nonlinearities in the system are key elements in guaranteeing the functionality and quality of the simulation. A detailed examination of these aspects of digital analysis lies outside of the scope of this book; for more information, see Allaire (2012). Moreover, diffusion or reaction terms, which are spatially and/or temporally heterogeneous, must undergo pre-processing before they can be used in a simulator. Finally, simulator output must often undergo post-processing to permit direct comparison with observation data (aggregation across a sub-domain of the zone of study, evaluation of the solution for specific places and at specific times, etc.). All of these elements mean that specific numerical expertise is required in order to develop a simulator. PDE simulation programs do exist, but these are not always easy to use in the context of a mechanistic-statistical study due to the number of couplings that would be required.

4.3. Estimation

4.3.1. *Estimation principle*

The models used to represent population dynamics generally contain unknowns, which may take the form of parameters (e.g. capacity of the habitat, diffusion parameter, and so on), functions (i.e. parameters of infinite dimension, e.g. reproductive rate as a function of spatial position), or latent (hidden) processes or variables (e.g. non-observed predator numbers in a situation featuring predator–prey dynamics). In this chapter, we have chosen to focus on the estimation of parameters and the inference of latent processes; for cases requiring the estimation of functions, readers may wish to consult the literature on semi-parametric and non-parametric models (Ruppert *et al.* 2003). The latter book also offers an introduction to parameter estimation (for theoretical foundations, see Dacunha-Castelle and Duflo 1994), and an introduction to parameter estimation in a Bayesian context can be found in Albert *et al.* (2015). Excellent works also exist on the subject of mixed and hierarchical models (McCulloch and Searle 2001; Albert *et al.* 2015) in relation to latent variables and processes.

4.3.2. *Parameter estimation*

Let θ be the unknown parameters in a model $\mathcal{M}_\theta$, and let these parameters belong to a space Θ, which is not reduced to a singleton. To estimate the parameters of the model using data, we must evaluate the $\theta_0 \in \Theta$ under which the existing data were obtained, given that the model that generated these data is contained within the model class $\{\mathcal{M}_\theta : \theta \in \Theta\}$. In many real-world situations, this approach consists of estimating θ in the context of a regression (often nonlinear), for example of the form:

$$y = \mathcal{M}_\theta(x) + \epsilon \qquad [4.17]$$

where y is the response variable, x is an explanatory variable, $\mathcal{M}_\theta$ is a deterministic function of x and ϵ is a noise. Once model [4.17] has been specified, the aim is to infer θ using data which, in the simplest case, consist of repeated and independent observations of (x, y), that is, a sample $\{(x_i, y_i) : i = 1, \ldots, I\}$.

The ordinary least squares estimator: the most "intuitive" approach is to determine the value of θ by minimizing, the sum of squared deviations between the observations y_i and their predicted values $\hat{y}_i = \mathcal{M}_\theta(x_i)$:

$$\hat{\theta} = \text{argmin}_{\theta \in \Theta} \sum_{i=1}^{I} \{y_i - \mathcal{M}_\theta(x_i)\}^2$$

In the general case, this minimization must be carried out using an iterative numerical algorithm (Gauss–Newton, Nelder–Mead, simulated annealing, etc.), but exact expressions are available in some cases. Concerning the uncertainty of $\hat{\theta}$, if the noise ϵ is zero, the inference is exact ($\mathcal{M}_{\hat{\theta}}$ is an interpolator). If this is not the case, then the structure of the noise and the random element associated with the drawn sample must be taken into account. In simple cases, estimation by least squares offers a means of describing the law, or at least the variance, of $\hat{\theta}$ in an exact or asymptotic manner (i.e. when I tends toward infinity). The two methods presented below, which are two of the major paradigms of modern statistics, offer a more generic, elegant approach.

Maximum likelihood estimation (MLE): in the case of regression, a distributional hypothesis is required for noise. For example, in $Y_i = \mathcal{M}_\theta(x_i) + \epsilon_i$, the ϵ_i are independent and identically distributed following the Normal law $(0, \sigma^2)$ where $\sigma \geq 0$, and Y_i is the random variable taking the value y_i through sampling. Using such a hypothesis, we must write the joint law of Y_{is} given x_{is}, which is dependent on the parameter θ. We must then maximize this function (the likelihood) with respect to θ.

In the case of a Gaussian noise, the conditional law of the random variable Y is thus:

$$Y \mid X = x \sim \text{Normal}(\mathcal{M}_\theta(x), \sigma^2). \quad [4.18]$$

The probability density of Y given $X = x$ is thus:

$$y \mapsto f_{Y|X}(y|x; \theta, \sigma) = \frac{1}{\sigma\sqrt{2\pi}} \exp\left(-\frac{(y - \mathcal{M}_\theta(x))^2}{2\sigma^2}\right).$$

Given the hypothesis that ϵ_i are independent, the likelihood is written as follows:

$$L(\theta; (y_i, x_i) : i = 1, \ldots, I, \sigma) = \prod_{i=1}^{I} f_{Y|X}(y_i|x_i; \theta, \sigma). \quad [4.19]$$

The maximum likelihood estimator is

$$\hat{\theta} = \text{argmax}_{\theta \in \Theta} L(\theta; (y_i, x_i) : i = 1, \ldots, I, \sigma).$$

In certain cases, there is an analytical expression of $\hat{\theta}$. Otherwise, a maximization algorithm, such as those mentioned earlier, is used. For numerical reasons, it is the log-likelihood that is generally maximized. If, in these specific cases, the uncertainty of $\hat{\theta}$ can be characterized specifically, in the general case and in accordance with regularity hypotheses, we obtain a result of asymptotic normality:

$$\sqrt{I}(\hat{\theta} - \theta) \rightarrow_d \text{Normal}(0, \mathcal{I}_\theta^{-1}),$$

where $\mathcal{I}_\theta$ is the Fisher information matrix.

Bayesian estimation: in this context, parameter θ is considered as a random variable for which prior knowledge (expertise) is available, modeled by a probability density $\theta \mapsto \pi(\theta)$ defined on Θ. The posterior distribution of θ combines prior information with the information drawn from data. In the continuous case, it is expressed in the form

$$f(\theta \mid \text{data}) = \frac{f(\text{data} \mid \theta)\pi(\theta)}{f(\text{data})},$$

where $f(\text{data} \mid \theta)$ is the likelihood of the model and $f(\text{data}) = \int_\Theta f(\text{data} \mid \theta)\pi(\theta)d\theta$. For example, in the case of model [4.18] above, $f(\text{data} \mid \theta) = \prod_{i=1}^{I} f_{Y|X}(y_i|x_i; \theta, \sigma)$. The posterior evaluation of the distribution provides us with pointwise estimations of θ (posterior mode, mean or median), and also allows us to evaluate the variability of θ, taking account of prior knowledge and of the information obtained from data.

For specific expressions of likelihood and prior distribution, the posterior distribution corresponds to a usual distribution. In more complex cases, an algorithm may be used to sample the posterior distribution, and the drawn sample (ideally large) is used to evaluate the characteristics of the posterior distribution (mode, mean, median, quantiles, etc.) numerically. The most commonly used algorithms in these cases are those based on importance sampling, MCMCs with a Gibbs or Metropolis–Hastings sampler, and their derivatives or extensions.

Other estimation methods may be used to work around potential problems encountered when using the approaches described above. Notable examples include the method of moments (MM), or approximate Bayesian computation (ABC), which used summarized statistics in the place of raw data; another approach involves estimating pseudo-likelihood or quasi-likelihood, whereby dependency relations between data elements or distributional hypotheses may be ignored.

4.3.3. *Estimation of latent processes*

Latent processes are integrated into models with a hierarchical structure (state–space models). Typically, taking equation [4.17], the response variable y is modeled conditional to a process $\mathbf{Z}$:

$$y = \mathcal{M}_\theta(x, \mathbf{Z}) + \epsilon, \qquad [4.20]$$

where $\mathbf{Z}$ is generated by a stochastic model parameterized by θ. More generically, ϵ may even be dependent on θ, x and $\mathbf{Z}$. In the general case, the chosen model structure has limited dependency relations between variables and parameters. For example, in many cases, $\mathcal{M}_\theta$ is only parameterized by a subset of θ, say θ_1, and $\mathbf{Z}$ is modeled as a function of the remaining components of θ, say θ_2; y does not depend on the whole process $\mathbf{Z}$, but on only one component, corresponding to elements in x. Typically, if x contains spatial coordinates s, a time t and environmental covariates, and if $\mathbf{Z}$ is a process indexed by space and time (with $Z(s,t)$ denoting the value of $\mathbf{Z}$ at (s,t)), then [4.20] becomes

$$y = \mathcal{M}_{\theta_1}(x, Z(s,t)) + \epsilon.$$

We shall now consider a real-world example, which will be used in a slightly different form in section 4.4.2. Take y_i an observed number of individuals at a site with coordinates $s_i \in \Omega \subset \mathbb{R}^2$ and at a time $t_i \in \mathbb{R}$. The corresponding random variable, noted Y_i, obeys the following conditional Poisson distribution:

$$Y_i \mid Z(s_i, t_i) \sim \text{Poisson}(\lambda_{\theta_1}(x) \exp\{Z(s_i, t_i)\}), \qquad [4.21]$$

where λ is a positive deterministic function parameterized by θ_1 and dependent on space, time, and any environmental covariates. Given the values of $Z(s_i, t_i)$, $i = 1, \ldots, I$, the values of Y_i are independent. $\mathbf{Z}$ is a Gaussian spatio-temporal field parameterized by θ_2, that is, for any finite set of points in the space-time domain, the vector of the values taken by $\mathbf{Z}$ at these points is Gaussian with a mean and variance–covariance matrix parameterized by θ_2. For any strictly positive integer n and any n points in the space-time domain,

$$\{Z(s,t); (s,t) \in (\Omega \times \mathbb{R})^n\} \sim \text{Normal}(\mu_{\theta_2}, \Sigma_{\theta_2}), \qquad [4.22]$$

where the values of μ_{θ_2} and Σ_{θ_2} depend on the points (s,t) under consideration. In a model of this type, the values of $Z(s_i, t_i)$ are dependent but the values of Y_i given $Z(s_i, t_i)$ are conditionally independent. This simplifies the computation of likelihood, and, consequently, inference.

The likelihood of the model must be written in both maximum likelihood and Bayesian estimation approaches. In the general case, this takes the following form:

$$L(\theta; (y_i, x_i) : i = 1, \ldots, I) = \int f(\mathbf{Y} \mid \mathbf{x}, \mathbf{Z}; \theta) d\nu_\theta(\mathbf{Z}),$$

where $\mathbf{Y}$ is the set of Y_i, $\mathbf{x}$ is the set of x_i, $f(\mathbf{Y} \mid \mathbf{x}, \mathbf{Z}; \theta)$ is the joint conditional probability measure of Y_i, ν_θ is the probability measure of $\mathbf{Z}$ and integration is carried out with respect to $\mathbf{Z}$. Taking Y_i to be conditionally independent given $\mathbf{Z}$, then $f(\mathbf{Y} \mid \mathbf{x}, \mathbf{Z}; \theta)$ can be written as a product of I terms, in the same way as the product of equation [4.19] (e.g. $\prod_{i=1}^{I} f(Y_i \mid x_i, Z(s_i, t_i); \theta)$). However, if $\mathbf{Z}$ is actually a process incorporating the dependency characteristics of a phenomenon (and not simply a collection of independent variables), then the likelihood of the model remains a multiple integral (typically of size I), which is difficult to calculate. Different algorithms may be used in this case, such as EM (expectation–maximization), MCEM (Monte Carlo expectation–maximization), MCMC or importance sampling algorithms. In these algorithms, the state of $\mathbf{Z}$ (or at least of the components of $\mathbf{Z}$ used in writing the likelihood: e.g. the $Z(s_i, t_i)$ in the case of the model using equations [4.21]–[4.22]) is updated, alongside parameter values, with each iteration.

4.3.4. *Mechanistic-statistical models*

Here, the aim is to construct a model that connects a mechanistic representation of the underlying processes with data derived from these processes. To do this, we combine a sub-model of the observation process – a data model – with a mechanistic sub-model of the phenomenon in question – a process model.

In the case of a mechanistic-statistical model without latent processes (equation [4.17]), the mechanistic sub-model is $\mathcal{M}_\theta(x)$ and corresponds, for example, to an ODE or a PDE. The observation process corresponds to the noise ϵ.

In the case of a mechanistic-statistical model with latent processes (equation [4.20]), the mechanistic sub-model may correspond to $\mathcal{M}_\theta(x, \mathbf{Z})$, $\mathbf{Z}$ or both. Evidently, situations arise where $\mathcal{M}_\theta(x, \mathbf{Z})$ and $Z(x)$ coincide (in this case, $\mathbf{Z}$ is indexed by x or some of the components of x). If $\mathcal{M}_\theta(x, \mathbf{Z})$ forms part of the mechanistic sub-model, the observation process corresponds to ϵ. If the mechanistic sub-model is represented by $\mathbf{Z}$ alone, then the observation process corresponds to $\mathcal{M}_\theta(x, \mathbf{Z}) + \epsilon$. As we have seen, the mechanistic sub-model may take the form $\mathcal{M}_\theta(x, \mathbf{Z})$, and in this case, the term $\mathbf{Z}$ may be interpreted as a (random) function that conditions the mechanistic model. Thus, $\mathbf{Z}$ may be used to represent a non-observed environmental model that conditions the dynamics of a population. Consider the following example: $\mathcal{M}_\theta(x, \mathbf{Z})$ is the solution of a PDE giving the density of a population of interest, of which one parameter (in this case, $\mathbf{Z}$) varies over time and space and corresponds to a Gaussian process representing an environmental variable, or to a stochastic dynamic process giving the density of a predatory population. In other cases, $\mathbf{Z}$ may be a birth-death process, or, more generally, an SDE giving the size of a population, and $\mathcal{M}_\theta(x, \mathbf{Z})$ reflects the specificities of a relatively complex observation process (e.g. the observation may relate to cumulative damage over space and time caused by the presence of a given population).

The parameter estimation and latent process reconstruction approaches presented in sections 4.3.2 and 4.3.3 may be applied in all of these cases. The examples given below concern real situations, showing how the relatively generic frameworks described here can be adapted to the specific characteristics of real case studies.

4.4. Examples

4.4.1. *The COVID-19 epidemic in France*

Context: the COVID-19 epidemic was first detected in December 2019 in the Hubei province of China. The disease then propagated around the world, prompting the WHO to declare a pandemic on March 11, 2020. The first cases were officially detected in France on January 24. The infection fatality ratio (IFR), defined as the number of deaths over the number of infected individuals, is an important quantity in calculating the expected number of victims by the end of the epidemic, once a certain proportion of the population has been infected. While the data concerning the number of deaths due to COVID-19 are likely to be precise, the real number of infected individuals in the population is unknown, largely due to the relatively low testing rate in France in the early stages of the epidemic. The IFR cannot, therefore, be calculated directly. Using the first set of data available for France (up to March 17), we aim to (1) calculate the IFR for France, (2) estimate the number of individuals infected with COVID-19 in France, and (3) calculate the basic reproduction rate R_0. This example is taken from a study carried out by Roques *et al.* (2020b).

Data: the data used here are for COVID-19 testing in France from January 22, 2020 up to March 17, 2020. These data give the number of positive cases and deaths per day (Dong *et al.* 2020), along with the number of tests carried out during this period. Official data on COVID-19 deaths in France only cover deaths in hospitals. Data from the Grand Est region indicate an additional 570 deaths in nursing homes (EHPAD), and these data must be added to the official total (1,015 deaths as of March 31).

Mechanistic model: we shall use the SIR model described by the equation system [4.10], beginning at an estimated time $t = t_0$ that should correspond, approximately, to the data of introduction (t_0 may be considered as an *equivalent* introduction date in a dynamic situation in which a single introduction determines the beginning of the epidemic). The initial value $S(t_0) = 67 \cdot 10^6$ represents the total population of France; $I(t_0) = 1$ and $R(t_0) = 0$. Parameter β is fixed using information concerning the period of contagion ($\beta = 1/10$, corresponding to a duration of 10 days). Parameter α will be estimated. Let $D(t)$ be the number of deaths due to the epidemic. The impact of compartment $D(t)$ on the dynamics of the SIR system and on the total population will be ignored. The dynamics of $D(t)$ are dependent on $I(t)$ via the differential equation:

$$D'(t) = \gamma(t)\, I(t), \qquad [4.23]$$

where $\gamma(t)$ is the death rate among infected individuals.

Model [4.10] can be solved analytically by changing the time variable, requiring numerical integration. In practice, this method is not efficient, and a standard numerical solution using the Matlab solver *ode45* is preferable.

Observation model: the number of positive test cases on day t, denoted as $\hat{I}_t$, is presumed to follow a binomial distribution conditional on the number of tests n_t on day t and on the probability p_t of a positive test within this sample:

$$\hat{I}_t \sim Binomial(n_t, p_t). \qquad [4.24]$$

The tested population comprises a proportion of infected individuals and a proportion of susceptible individuals: $n_t = \tau_1(t)\, I(t) + \tau_2(t)\, S(t)$. Thus,

$$p_t = \frac{\sigma\, \tau_1(t)\, I(t)}{\tau_1(t)\, I(t) + \tau_2(t)\, S(t)} = \frac{\sigma\, I(t)}{I(t) + \kappa_t\, S(t)}$$

with $\kappa_t := \tau_2(t)/\tau_1(t)$: the ratio between the probability of being tested conditional on the fact of being of type S and the probability of being tested conditional on the fact of being of type I. The ratio κ will be considered independent of t for the period in question (start of the epidemic). The coefficient σ corresponds to the sensitivity of the test. In most cases, RT-PCR tests were used, and existing data indicate a sensitivity of around 70% ($\sigma = 0.7$).

Inference: the parameters to estimate are α, κ and t_0. Taking the increments $\hat{I}(t)$ to be independent conditional on the process, and for a given n_t, the likelihood of the parameters is:

$$\mathcal{L}(\alpha, t_0, \kappa) := P(\{\hat{I}_t\}|\alpha, t_0, \kappa) = \prod_{t=t_i}^{t_f} \frac{n_t!}{(\hat{I}_t)!(n_t - \hat{I}_t)!} p_t^{\hat{I}_t} (1 - p_t)^{n_t - \hat{I}_t}$$

with t_i the date of the initial observation and t_f the date of the final observation. In this expression, $\mathcal{L}(\alpha, t_0, \kappa)$ is dependent on α, t_0, κ via p_t. The posterior distribution of (α, t_0, κ) is calculated using a Bayesian method, using prior uniform distributions $\alpha \in (0, 1)$, $t_0 \in (1, 50)$, (January 1–February 19) and $\kappa \in (0, 1)$. This is done using a Metropolis-Hastings (MCMC) algorithm with four independent chains, each made up of 10^6 iterations.

Results: the number of observations, in this case the cumulative number of positive cases $\Sigma_t := \Sigma_{i=1,\dots,t}\hat{I}_i$, is compared with the expectation of the observation model associated with the posterior mode: $n_t\, p_t^*$ (expectation of a binomial distribution), with

$$p_t^* = \frac{\sigma\, I^*(t)}{I^*(t) + \kappa^*\, S^*(t)}$$

and where $I^*(t)$, $S^*(t)$ are the solutions of the system [4.10] associated with the posterior mode. The results in Figure 4.1 show that these results are well fitted to the data.

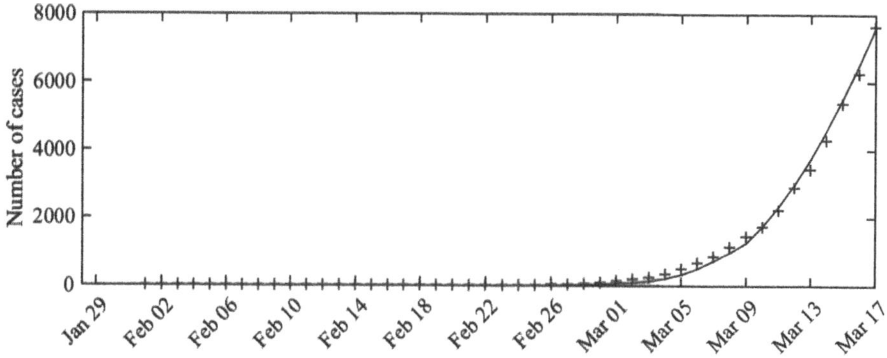

Figure 4.1. *Expectation of the total number of cases associated with the posterior mode versus number of cases actually detected (cumulative). The curve shows the expectation $n_t\, p_t^*$ obtained from the observation model, while the crosses correspond to the data (cumulative values of Σ_t). For a color version of this figure, see www.iste.co.uk/peyrard/ecology.zip*

The joint posterior distributions of the three pairs of parameters (α, κ), (t_0, α) and (t_0, κ) are shown in Figure 4.2. Note that these distributions are very different from the prior uniform distribution (indicating that the data are sufficiently informative) and that the support is sufficiently far from the boundaries of the *a priori* intervals (indicating that the intervals used in defining the prior distribution were sufficiently large). By simulating [4.10] using the prior parameter distributions, we can calculate the distribution of the latent variables I and R. This gives us a factor of 8 between observed and real cases.

In the case of model [4.10], R_0 can be calculated easily using the formula $R_0 = \alpha/\beta$ (Murray 2002). Using the marginal prior distribution of α, we obtain $R_0 = 3.2$ (95% CI: 3.1–3.3). Note that R_0 corresponds to the average number of people infected by an individual of type I in an entirely susceptible population. The epidemic will only propagate if $R_0 > 1$. Using relation [4.23], the estimated distribution of $I(t)$ and the mortality data $D(t)$, the distribution of the parameter $\gamma(t)$ for each date can be calculated. The IFR corresponds to the proportion of infected individuals who die from the disease, that is, $IFR_t := \gamma(t)/(\gamma(t) + \beta)$. This gives us an IFR of 0.5% (95% CI: 0.3–0.8) for March 17, excluding the nursing home data. Using local data for the Grand Est region, we see that this IFR should be multiplied by a factor of around $(1015 + 570)/1015 \approx 1.6$, giving an adjusted IFR of 0.8% (95% CI: 0.45–1.25) for the French population as a whole, including nursing home residents.

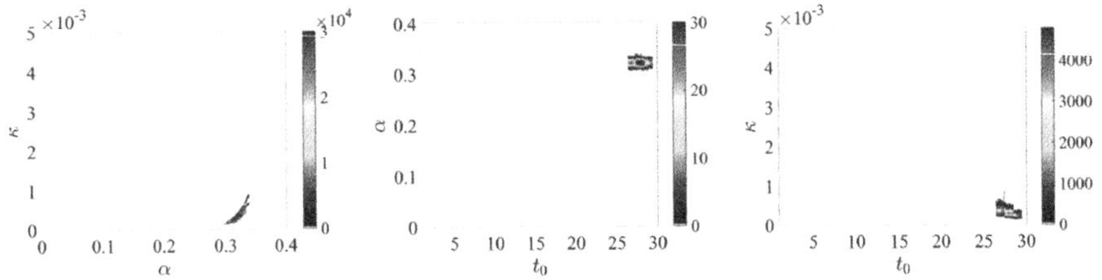

Figure 4.2. *Joint posterior distributions of couples (α, κ), (t_0, α) and (t_0, κ). For a color version of this figure, see www.iste.co.uk/peyrard/ecology.zip*

4.4.2. *Wolf* (Canis lupus) *colonization in southeastern France*

Context: most species distribution models are based on a regressive approach using presence indicators and as a function of environmental covariates, and thus implicitly suppose that all favorable zones are colonized. However, this hypothesis can result in a bias in predictions concerning the future presence of a species, notably in the case of species with strong colonization dynamics. In this section, the use of mechanistic models will be illustrated using the case of the recolonization of southeastern France by gray wolves (*Canis lupus*) between 2007 and 2015. This example is taken from a study carried out by Louvrier *et al.* (2020).

Data: detection and non-detection data were collected by a network of professional and non-professional observers, looking for signs that wolves were present (Duchamp *et al.* 2012). Signs were then filtered in order to avoid false positives. Detections mostly occurred in winter, from December to March. We shall divide the data into four detection events (December, January, February and March), and the population will be considered to be closed during this period. Our study is limited to southeastern France and to the period 2007–2015 (Figure 4.3). The presence–absence of wolves is defined using a network of regular 10 x 10 km cells covering the zone of study.

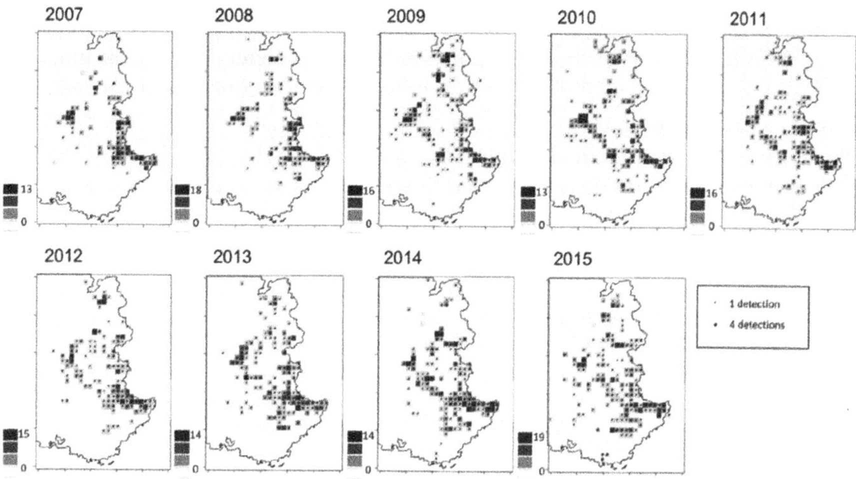

Figure 4.3. *Map of wolf detections in southeastern France (black dots) and the abundance predicted by the model (graduated colors) for each 100 km² site. For a color version of this figure, see www.iste.co.uk/peyrard/ecology.zip*

Mechanistic model: Let $N_{i,t}$ be the random latent variable giving the abundance of wolves at a site i in year t, such that:

$$\begin{cases} N_{i,t} \sim \text{Poisson}\big(\lambda(i,t)\epsilon_{i,t}\big), \\ \log(\epsilon_{i,t}) \sim \text{N}(0,\sigma) \end{cases} \qquad [4.25]$$

Variable $\lambda(i,t)$ gives the theoretical abundance at site i in year t, while $\epsilon_{i,t}$ is used to consider overdispersion around this value. The theoretical abundance is calculated using the solution to a reaction–diffusion equation describing the spatiotemporal dynamics of wolf colonization:

$$\begin{cases} \lambda(i,t) = \displaystyle\int_{B_i} \nu(t,x)dx, \\ \dfrac{\partial \nu(t,x)}{\partial t} = \Delta\big(d(x)\nu(t,x)\big) + f(x,\nu(t,x)), \end{cases} \qquad [4.26]$$

with

$$f(x,\nu(t,x)) = \begin{cases} r(x)\nu(t,x)(1 - \frac{\nu(t,x)}{K}) \text{ when } r(x) > 0, \\ r(x)\nu(t,x) \text{ when } r(x) \leq 0. \end{cases} \qquad [4.27]$$

In this case, we consider that the diffusion coefficient and the intrinsic growth rate depend on the spatial position x. More precisely, these parameters depend, respectively, on human population density and forest cover via a linear and quadratic effect. In order to respect the constraints on these parameters, a log link function is used for the diffusion coefficient, with a logistic function constrained between 0 and 2 for the growth rate. In order to calculate $\nu(t,x)$, equation [4.26] must be calculated. This is done using the method of lines (Schiesser 1991; Chow 2003) , which consists of approximating the PDE via a system of ODEs, discretizing the space as a regular grid in order to apply classic algorithms. Equation [4.26] can then be approximated using the following ODE system:

$$\begin{cases} \lambda(i,t) = \sum_k \sum_l \mathcal{A}(B_i \cap c_{s(k,l)}) u(s_{k,l}, t), \\ \dot{U}_t = R \times U_t \left(\mathbb{1}_{R<0} + \left(1 - \mathbb{1}_{R<0}\right)\left(1 - \frac{U_t}{K}\right)\right) + MU_t, \end{cases} \quad [4.28]$$

where $\mathbb{1}_{R<0}$ is the vector indicating negative elements of R and $\mathcal{A}(B_i \cap c_{s(k,l)})$ is the surface where cell $s(k,l)$ in the grid and the domain B_i in which the observation took place intersect. $\times$ denotes the term by term product. $U_t^T = [u(1,t), ..., u(C_s,t)]$, where C_s is the total number of cells, is the vector of densities in each cell in the grid, and $R^T = [\bar{r}(1), ..., \bar{r}(C_s)]$ is the vector of mean growth rates for each cell. M is the propagation matrix approximating the diffusion operator.

Observation model: let y_{ijt} be the random variable, with a value of 1 if at least one individual was detected at site $i = 1, ..., K$ during capture event $j = 1, ..., J$ in year $t = 1, ..., T$, and 0 otherwise. We obtain

$$\sum_{j=1}^{J} y_{ijt} \sim \text{Binomial}\left(J, p_{it}\right).$$

The probability p_{it} that at least one individual will be detected during a capture event depends on the individual probability of detection q_{it} and on the abundance $N_{i,t}$ of wolves at site i in year t via the relationship

$$p_{it} = 1 - \left(1 - q_{it}\right)^{N_{it}},$$

in which abundance is obtained using the mechanistic model. Variations in the individual probability of detection are taken into account via a log-linear relation between q, the sampling effort Eff and road density rD:

$$\text{logit}\left(q_{it}\right) = \beta_0 + \beta_1 \text{Eff}_{it} + \beta_2 \text{rD}_i.$$

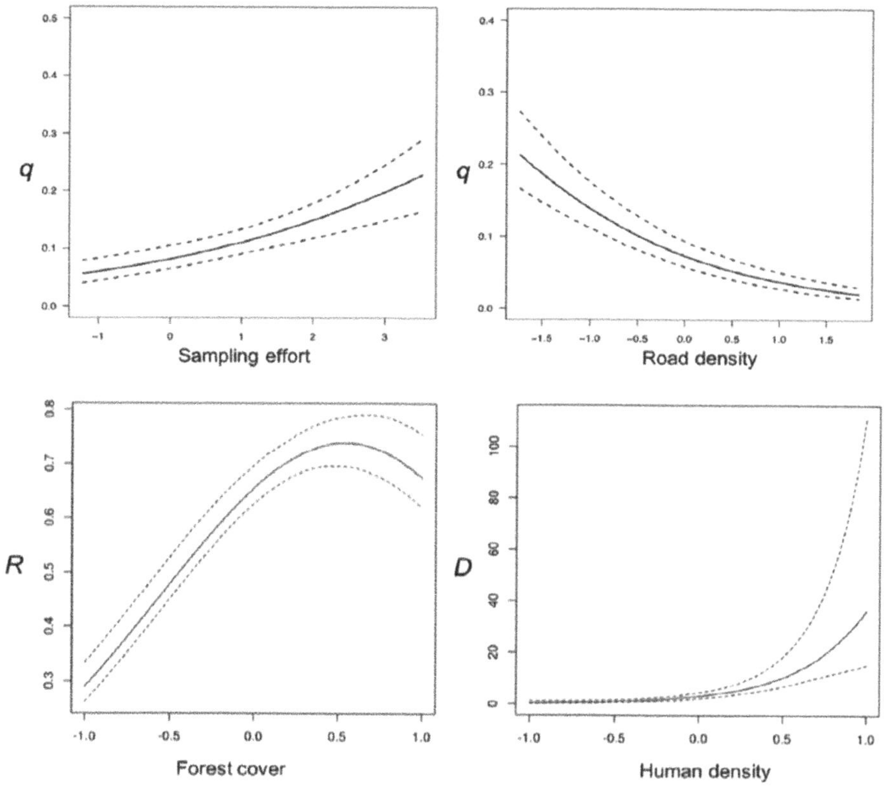

Figure 4.4. *Estimated response curves. Estimated relations between individual detectability and sampling effort (top left) or road density (top right), between growth rate and forest coverage (bottom left), and between the diffusion coefficient and human population density (bottom right). For a color version of this figure, see www.iste.co.uk/peyrard/ecology.zip*

Inference: inference was carried out in a Bayesian framework using the *mecastat* module (Rey *et al.* 2018) in JAGS. Gaussian prior distributions with a mean of 0 and variance of 1 were defined for all parameters, except for K, for which a logistical link function was used, limited to an interval between 0 and 0.2. The mechanistic model defined in system [4.28] was solved using the *deSolve* package (Soetaert *et al.* 2010) in R.

Results: overall, the abundances predicted by the model are a good reflection of the spatiotemporal trends observed in the detection data (Figure 4.3). The probability of detection increases with sampling effort, and decreases as road density increases (Figure 4.4). Furthermore, the growth rate is seen to increase with forest coverage. This effect is certainly due to a greater abundance of prey. Finally, human population density increases the diffusion coefficient, reducing the abundance of wolves in highly populated zones.

The model was fitted using data from 2007 to 2015, and was used to predict wolf presence in 2016. Figure 4.5 shows a comparison of predicted and observed presence for this year. Wolves were detected at many of the sites where the predicted probability of occupation was high. Out of 137 sites at which wolves were detected in 2016, only 10, all in the southwestern part of the study zone, had a low predicted probability of occupation. On the other hand, the model predicted that wolves would be present at a higher number of sites where occupation was not detected in 2016. This may be due to non-detection of actual presence, but may also be explained by the diffusive aspect of dispersion, whereby the establishment of packs on the edges of the colonization area will not be detected immediately.

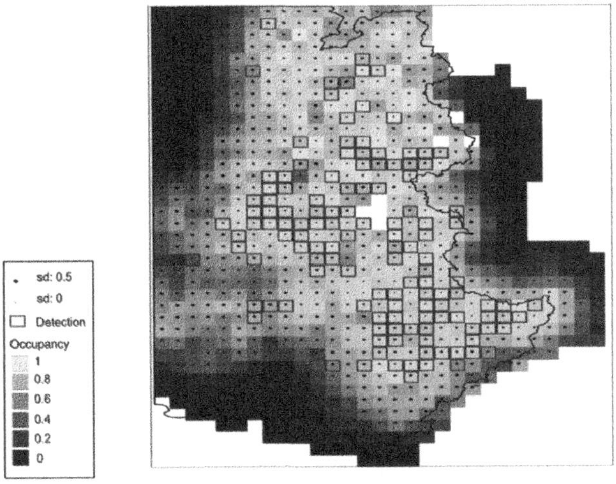

Figure 4.5. *Predicted occupation probability map for 2016, obtained using the model fitted to data from 2007 to 2015. The blue squares represent sites were wolves were actually detected in 2016. The black dots represent the uncertainty of the prediction. For a color version of this figure, see www.iste.co.uk/peyrard/ecology.zip*

4.4.3. *Estimating dates and locations of the introduction of invasive strains of watermelon mosaic virus*

Context: determining the place and time of introduction of exogenous species is a key issue in invasion biology; this information is key to understanding the biotic and abiotic conditions which promote the introduction and establishment of invasive species. However, in many cases, biological invasions go undetected until several years after the initial introduction. For this reason, the development of methods for estimating the date and location at which a species first appeared is key. Any model used for this purpose must take account of lifecycle parameters, as the estimation of these parameters also depends on the distance and duration separating the actual introduction and observation. In this example, a reaction–diffusion model will be used to describe the invasion dynamics of multiple species or genotypes in competition with an established resident population. This example is taken from a study carried out by Roques *et al.* (2020a).

Data: the data in this case consist of a set of observations describing the invasion of a landscape, initially occupied by a resident strain of watermelon mosaic virus (WMV, Potyvirus genus), which is particularly widespread in cucurbit crops, by four invasive variants. The area of study is located in southeastern France. The presence of WMV in plants was monitored from 2004 to 2008, with over 200 specimens collected and analyzed each year. Each observation for a given date and given location corresponds to the number of plants infected with each viral strain. The quality of the habitat in the area of study was approximated based on the proportion of plants susceptible to WMV in each cell in a raster, with a resolution of $1.4\times$ 1.4 km^2. This proportion was normalized to give the density of host plants in each cell x, $z(x)$, such that $z(x) = 0$ if no host plants were present in cell x and $z(x) = 1$ if the proportion of host plants was maximal (Figure 4.6).

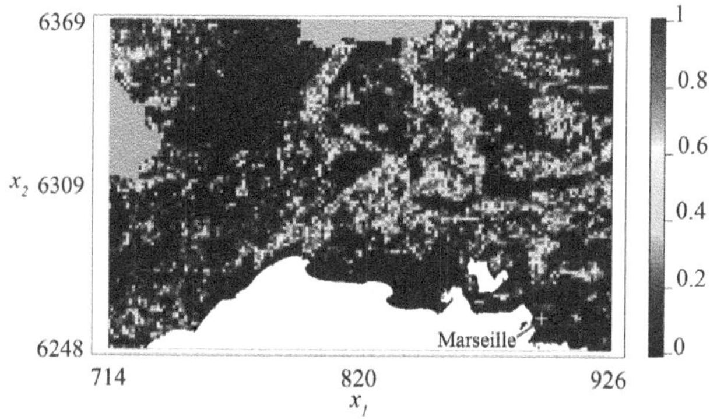

Figure 4.6. *Proportion of plants susceptible to WMV across the area of study. For a color version of this figure, see www.iste.co.uk/peyrard/ecology.zip*

Mechanistic model: let $C^n(t, x)$ and $E_k^n(t, x)$ be the densities of the resident (classic) strain and invasive (or mutant) variants k, respectively, at position x, time t and in year n. The intra-annual dynamics describe the dispersion and growth in the populations of the five competing viral strains over the course of the epidemic season, using the following model of neutral competition with diffusion:

$$\begin{cases} \partial_t C^n(t,x) = D\Delta C^n + rC^n\Big(z(x) - C^n - \sum_{i=1}^{4} E_i^n(t,x)\Big), \\ \partial_t E_k^n(t,x) = D\Delta E_k^n + rE_k^n\Big(z(x) - C^n - \sum_{i=1}^{4} E_i^n(t,x)\Big), \end{cases}$$

for $t \in [0, t_f]$ and for all invasive variants present during the year in question, that is for invasive variants E_k such that $n \geq n_k$ where n_k is the year of introduction of variant k. The boundary conditions are reflecting. To limit the number of parameters, and based on present knowledge of the virus, the diffusion, competition and growth coefficients for each variant are identical at this stage in the process. The inter-annual dynamics describe the mortality of different variants over winter, when there are no susceptible plants in the landscape and when the virus is forced to survive in the wild. Population densities at time $t = 0$ in year n are linked to those in year $n-1$ at time t_f in the following manner:

$$\begin{cases} C^n(0,x) = (1 - m_C)C^{n-1}(t_f, x), \\ E_k^n(0,x) = (1 - m_E)E^{n-1}(t_f, x), \end{cases}$$

where m_C and m_E are the winter mortalities of the resident and mutant variants, respectively.

Observation model: the data correspond to the number of samples infected by the classic strain, C_i^{obs}, and by the different invasive variants, $E_{k,i}^{\text{obs}}$, at a given date and time, (t_i, x_i). The vector $\big(C_i^{\text{obs}}, E_{1,i}^{\text{obs}}, E_{2,i}^{\text{obs}}, E_{3,i}^{\text{obs}}, E_{4,i}^{\text{obs}}\big)$ follows a multinomial distribution $\mathcal{M}\big(V_i, \mathbf{p}_i\big)$ where $V_i = C_i^{\text{obs}} + \sum_k E_{k,i}^{\text{obs}}$ is the total number of samples infected by the virus. $\mathbf{p}_i = \big(p_i^C, p_i^{E_1}, p_i^{E_2}, p_i^{E_3}, p_i^{E_4}\big)$ is the vector of the respective populations of each variant in the viral population at time t_i and in location x_i. These proportions are given by the mechanistic model as follows:

$$p_i^C = \frac{C^n(t_i, x_i)}{C^n(t_i, x_i) + \sum_k E_k^n(t_i, x_i)}, \; p_i^{E_k} = \frac{E_k^n(t_i, x_i)}{C^n(t_i, x_i) + \sum_k E_k^n(t_i, x_i)}$$

Inference: due to the long calculation time and the possibility of obtaining multiple local optimums, parameters were inferred using maximum likelihood by means of a simulated annealing algorithm. This is an acceptance/rejection type algorithm that constructs a sequence of parameters, converging in probability toward the maximum likelihood estimator.

Biological parameter	D	r	m_c	m_e
Value	0.44 km^{2}day^{-1}	0.31 day^{-1}	0.5 year^{-1}	0 year^{-1}
Date of introduction	E_1	E_2	E_3	E_4
Value	1990	1990	1990	1995
Place of introduction	E_1	E_2	E_3	E_4
Value (Lambert 93, km)	(926, 6369)	(926, 6369)	(758, 6369)	(758, 6294)

Table 4.1. *Parameters estimated by maximum likelihood*

Results: the estimated parameters are shown in Table 4.1. Three of the invasive variants appear to have been introduced in the same year, while invasive variant 4 appeared 5 years later. The relatively distant introduction sites explain the spatial partitioning of variants (Figure 4.7). In terms of biological parameters, the high winter mortality rate of the classic strain compared to the invasive variants shows that the latter have a strong competitive advantage. The estimated diffusion coefficient indicates that the virus moves 1.3 km/day on average. The estimated growth rate corresponds to an increase by a factor of 1.3 each day, in the absence of competition.

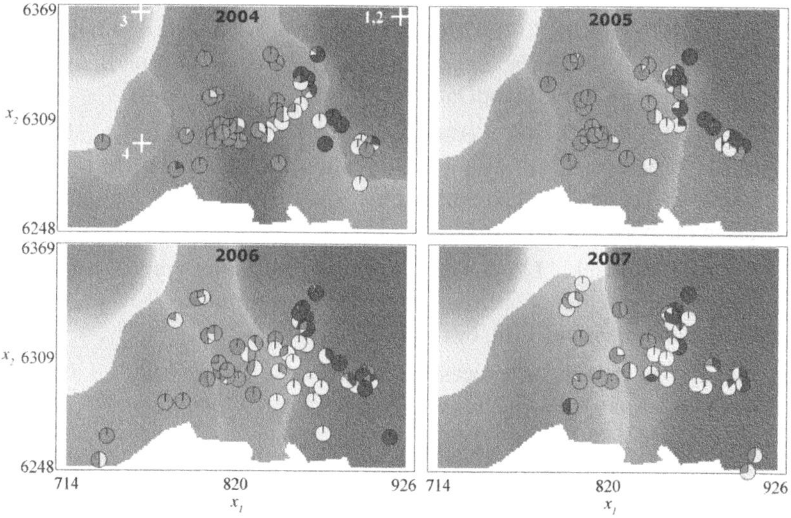

Figure 4.7. *Proportions of classic and invasive variants in a landscape: data and simulations. The classic strain is shown in red, while invasive variants are shown in light blue (E_1), dark blue (E_2), green (E_3) and purple (E_4). The proportions of each variant in the data are shown as pie charts. The white crosses indicate estimated sites of introduction. Colored zones show the spatio-temporal dynamics of the dominant strain, as simulated using parameters corresponding to the maximum likelihood estimator. For a color version of this figure, see www.iste.co.uk/peyrard/ecology.zip*

4.5. References

Albert, I., Ancelet, S., David, O., Denis, J.-B., Makowski, D., Parent, E., Rau, A., Soubeyrand, S. (2015). *Initiation à la statistique bayésienne : bases théoriques et applications en alimentation, environnement, épidémiologie et génétique*. Ellipses, Paris.

Allaire, G. (2012). *Analyse numérique et optimisation*. Éditions de l'École Polytechnique, 2nd edition [Online]. Available at: http://www.cmap.polytechnique.fr/~allaire/livre2.html.

Allee, W.C. (1931). *Animal Aggregations: A Study in General Sociology*. University of Chicago Press, Chicago.

Baake, E. and Wakolbinger, A. (2015). Feller's contributions to mathematical biology. arXiv preprint arXiv:1501.05278.

Chow, S.-N. (2003). Lattice dynamical systems. In *Dynamical Systems*, Macki, J.W., Zecca, P. (eds). Springer, Berlin.

Dacunha-Castelle, D. and Duflo, M. (1994). *Probabilités et statistiques, tome 1 : problèmes à temps fixe*. Masson, Paris.

Dong, E., Du, H., Gardner, L. (2020). An interactive web-based dashboard to track COVID-19 in real time. *The Lancet Infectious Diseases* [Online]. Available at: https://github.com/CSSEGISandData/COVID-19/.

Duchamp, C., Briaudet, P.-E., Leonard, Y., Moris, P., Bataille, A., Dahier, T., Delacour, G., Millisher, G., Miquel, C., Poillot, C., Marboutin, E. (2012). Wolf monitoring in France: A dual frame process to survey time- and space-related changes in the population. *Hystrix*, (23), 14–28.

Elith, J. and Leathwick, J.R. (2009). Species distribution models: Ecological explanation and prediction across space and time. *Annual Review of Ecology, Evolution and Systematic*, (40), 677–697.

Feller, W. (1951). Diffusion processes in genetics. *Proceedings of the Second Berkeley Symposium on Mathematical Statistics and Probability*. The Regents of the University of California.

Guisan, A. and Thuiller, W. (2005). Predicting species distribution: Offering more than simple habitat models. *Ecology Letters*, (8), 993–1009.

Hefley, T.J., Hooten, M.B., Russell, R.E., Walsh, D.P., Powell, J.A. (2017). When mechanism matters: Bayesian forecasting using models of ecological diffusion. *Ecology Letters*, (20), 640–650.

Louvrier, J., Papaïx, J., Duchamp, C., Gimenez, O. (2020). A mechanistic–statistical species distribution model to explain and forecast wolf (*Canis lupus*) colonization in south-eastern France. *Spatial Statistics*, 36, 100428.

McCulloch, C.E. and Searle, S.R. (2001). *Generalized, Linear and Mixed Models*. Wiley, New York.

Murray, J.D. (2002). *Mathematical Biology*, 3rd edition. Springer-Verlag, New York.

Papaïx, J., Burdon, J.J., Walker, E., Barrett, L.G., Thrall, P.H. (2021). Metapopulation structure predicts population dynamics in the *Cakile maritima–Alternaria brassicicola* host-pathogen interaction. *American Naturalist*, 197(2), E55–E71.

Rey, J.-F., Walker, E., Bonnefon, O., Papaïx, J. (2018). A JAGS package to fit mechanistic-statistical models to data [Online]. Available at: https://gitlab.paca.inra.fr/jfrey/jags-module.

Roques, L., Desbiez, C., Berthier, K., Soubeyrand, S., Walker, E., Klein, E.K., Garnier, J., Moury, B., Papaïx, J. (2020a). Emerging strains of watermelon mosaic virus in southeastern France: Model-based estimation of the dates and places of introduction. *bioRxiv* [Online]. Available at: https://doi.org/10.1101/2020.10.01.322693.

Roques, L., Klein, E., Papaïx, J., Sar, A., Soubeyrand, S. (2020b). Using early data to estimate the actual infection fatality ratio from COVID-19 in France. *MDPI Biology*, 9(5), 97.

Ruppert, D., Wand, M.P., Carroll, R.J. (2003). *Semiparametric Regression*. Cambridge University Press, Cambridge, UK.

Schiesser, W.E. (1991). *The Numerical Method of Lines: Integration of Partial Differential Equations*. Academic Press, San Diego.

Shigesada, N. and Kawasaki, K. (1997). *Biological Invasions: Theory and Practice*. Oxford University Press, Oxford.

Soetaert, K., Petzoldt, T., Setzer, R.W. (2010). Solving differential equations in R: Package deSolve. *Journal of Statistical Software*, 33(9), 1–25 [Online]. Available at: https://doi.org/10.18637/jss.v033.i09.

Soubeyrand, S., Laine, A.-L., Hanski, I., Penttinen, A. (2009). Spatiotemporal structure of host–pathogen interactions in a metapopulation. *The American Naturalist*, (174), 308–320.

Turchin, P. (1998). *Quantitative Analysis of Movement: Measuring and Modeling Population Redistribution in Animals and Plants*. Sinauer, Sunderland.

5

Using Coupled Hidden Markov Chains to Estimate Colonization and Seed Bank Survival in a Metapopulation of Annual Plants

Pierre-Olivier CHEPTOU[1], Stéphane CORDEAU[2], Sebastian LE COZ[3] and Nathalie PEYRARD[3]

[1] *CEFE UMR 5175, CNRS, University of Montpellier, Université Paul-Valery, France*

[2] *Agroécologie, AgroSup Dijon, INRAE, University of Bourgogne, Bourgogne Franche-Comté University, Dijon, France*

[3] *University of Toulouse, INRAE, UR MIAT, Castanet-Tolosan, France*

5.1. Introduction

The continued presence of a species in a landscape is the result of a balance between population extinction and recolonization (Levins *et al.* 1969). Local populations have limited lifespans due to changing environmental conditions, which make the habitat more or less favorable to a given population. Living organisms have adopted a range of ecological strategies to deal with changes in their environment, including the capacity to disperse their propagules. Furthermore, some organisms – notably plants – have also developed propagule dormancy strategies, delaying germination until favorable environmental conditions appear. In higher plants, seeds are used for both spatial dispersal and long-term survival via the creation of a bank of dormant seeds in the soil; these seeds may then germinate after several years, or even decades, of dormancy (Lewis 1973). Dormant seed banks are particularly common among annual plants (Vegis 1964) existing in a "disturbed" environment (deserts, agroecosystems, ruderal environments, etc.). In a context of environmental disturbance, the dynamics of annual plants at regional level are determined by their soil dormancy and spatial dispersal capacities. Furthermore, these traits condition the coexistence of species at community level (Leibold and Miller 2004). Nevertheless, it remains difficult to distinguish between the contributions of dormancy and spatial dispersal in the natural dynamics of plants. Simply observing a plant population within a field sample (at flowering stage, for example) does not tell us whether these plants were established by dispersal from a neighboring population, or germinated from a local dormant seed bank. The observation of adult plants alone may result in false conclusions concerning local extinctions or colonizations (Freckleton and Watkinson 2002; Freville *et al.* 2013).

The dynamics of plants exhibiting dormancy behaviors is typically a form of metapopulation dynamic (Levins *et al.* 1969), with local dynamic in each patch (e.g. a field) combined with dispersal between patches. From a modeling perspective, however, this is a metapopulation with hidden stage, since the state of the seed bank is usually unknown (the cost of acquisition is too high). In recent years, models of plant dynamics including dormancy have been developed using hidden Markov models (HMM), with the aim of producing a statistical estimation of key parameters in these dynamics. Examples include Lamy *et al.* (2013), Freville *et al.* (2013) and Manna *et al.* (2017), although the last two studies ignore the effect of seed setting by standing flora on the local seed bank. Other models include this effect, but do not include explicit spatial colonization, using a simpler propagule rain approach (Borgy *et al.* 2015; Pluntz *et al.* 2018). Furthermore, the observed and hidden variables in some of these models are considered to be of the presence–absence type, meaning that abundance or abundance class information, which is often present in the data, is lost. Nevertheless, these approaches clearly show the relevance of HMM techniques for modeling the dynamics of populations with a dormant stage, and for extracting information on plant dynamics based on visible flora. The model presented in Freville *et al.* (2013), for example, indicates whether or not a specific dynamic features a dormant stage. Pluntz *et al.* (2018) used an HMM to automatically construct homogeneous functional groups, consistent with current ecological knowledge, from available occurrence data.

In this chapter, we will present an HMM of local dynamics (within a patch) and regional dynamics (between patches) for an annual plant with a dormant stage, combining all of the advantages of existing models. Abundance class data will be taken into account, as will the effect of adult plants on the seed bank, alongside explicit spatial dispersal. This model may be seen as several HMM linked by the observations, hence the chosen name: multidimensional HMM with data feedback (MHMM-DF).

Weeds are plants that grow in cultivated land without having been planted there intentionally. Understanding the dormancy and colonization behaviors of these plants makes them easier to manage, and as such, it is of key importance in agronomy. If dormancy is the dominant factor in the dynamics of a species, local weed management approaches will focus on preventing seeds from entering the dormant state after falling (Gallandt 2006) or, on the contrary, on limiting the emergence of dormant seeds near cultivated crops in order to prevent simultaneous germination (Cordeau *et al.* 2017a,b). In cases where colonization plays a dominant role, management techniques will be applied at multi-patch level and will focus on eliminating plants before the seed sets, notably for species with high dispersal capacities, such as wind-borne species (Petit *et al.* 2011). In this chapter, we will use data from the experimental farm at Epoisses, France, to show how the MHMM-DF model highlights the relative roles played by different processes in plant dynamics (dormancy, dispersal and seed setting). We will then show how the effects of a crop planted in the same patch may be taken into account, inferring whether or not the germination period of this plant is also favorable to that of the weed species.

5.2. Metapopulation model for plants: introduction of a dormant state

5.2.1. *Dependency structure in the model*

Consider a metapopulation of a plant species over C patches. The population has two possible states: a dormant state, in which the seed bank is not observable, and a non-dormant state, with observable, emerged plants. The (main) mechanisms involved in the dynamics of this metapopulation are the survival of the seed bank, colonization, germination and seed setting. Two groups of variables are used to model these dynamics statistically, that is, for estimation purposes. The observed variables are the set of abundance classes of the non-dormant population in each patch c at each time step t, noted $Y_{c,t}$ (with values in $\Omega_Y = \{0, 1, \dots |\Omega_Y| - 1\}$). The hidden variables are the set of abundance classes of the dormant population in each patch c at each time step t, noted $Z_{c,t}$ (with values in $\Omega_Z = \{0, 1, \dots |\Omega_Z| - 1\}$). The dependency structure between these different variables should represent the different mechanisms listed above. This structure, in the case of the MHMM-DF model for annual plants and for two patches, is shown in Figure 5.1. The dormant population within each patch evolves following a Markovian dynamic (the future state only depends on the current state); however, unlike a classic HMM, $Z_{c,t}$ is not only dependent on $Z_{c,t-1}$, but also on $Y_{c,t}$ and on all $Y_{c',t}$ such that c' is sufficiently close to c for seed dispersal to occur from c' to c. There are thus three (groups of) variables influencing state $Z_{c,t}$: $Z_{c,t-1}$ via the survival of seeds in the soil, $Y_{c,t}$ via the production of new seeds which are added to the seed bank and $Y_{c',t}$ via colonization of the seed bank by seeds from other patches.

Finally, in this model, the dynamics of the non-dormant population are only dependent on the state of the local dormant population: $Y_{c,t}$ is only dependent on

$Z_{c,t-1}$ via a succession of processes (germination, growth of the plant, survival up to the seed production stage), crudely summarized here under the term "emergence".

The resulting dependency structure is illustrated in Figure 5.1. This structure is well suited to plants, as there are no arcs connecting dormant populations in two distinct patches. This corresponds to the fact that the dormant population (seeds in the seed bank) is immobile, and that dispersal involves seeds produced by the non-dormant population.

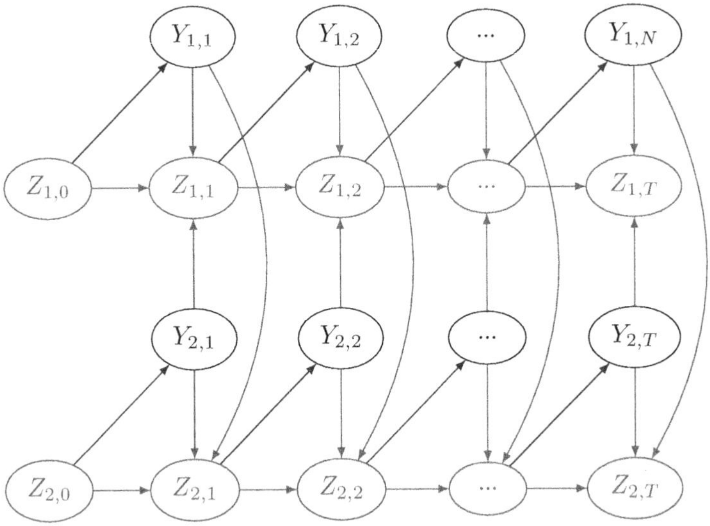

Figure 5.1. *Illustration of dependency relationships in an MHMM-DF. Case of two patches ($c = 2$), that is, two hidden chains (gray nodes). The blue lines reflect the probability of emission ϕ and the red lines show the probability of transition A of the state of the hidden chain. For each time step, each chain emits an observation (black nodes). The observations generated by all chains influence the next state of each of the hidden chains. The interaction between chains thus passes through observations. For a color version of this figure, see www.iste.co.uk/peyrard/ecology.zip*

However, this structure is not suitable for perennial plants due to the absence of arcs between $Y_{c,t-1}$ and $Y_{c,t}$. This notably implies that the non-dormant population cannot survive from t to $t+1$. If each time step represents a year, then this hypothesis is verified in the case of annual plants.

5.2.2. *Distributions defining the model*

An MHMM-DF model is fully defined by three distributions: the initial distribution, the emission law and the probability of transition. The first defines the initial distribution of all of the hidden states, that is, the abundance class of the seed bank in each of the C patches: $P(\boldsymbol{Z}_0^C = \boldsymbol{z}_0^C)$, where $\boldsymbol{Z}_0^C = (Z_{1,0}, \ldots, Z_{C,0})$ and $\boldsymbol{z}_0^C \in \Omega_Z^C$ is a realization[1] of $\boldsymbol{Z}_0^C$. The abundance classes are assumed to be independent and identically distributed variables, such that we can write $P(\boldsymbol{Z}_0^C = \boldsymbol{z}_0^C) = \prod_{c=1}^C \gamma_0(z_{c,0})$, with γ_0 a probability distribution on Ω_Z. The emission law then defines the probability of the abundance class of the non-dormant population at time t, conditional to the abundance class of the seed bank at time $t-1$. This is denoted as $\phi(z_{c,t-1}, y_{c,t}) = P(Y_{c,t} = y_{c,t} | Z_{c,t-1} = z_{c,t-1})$, and corresponds to the emergence process. Finally, the transition probability of a hidden chain c is the probability of the abundance class $Z_{c,t}$ of the seed bank at time t, conditional on the state of the seed bank at the previous time, $Z_{c,t-1}$, on the state of local standing flora $Y_{c,t}$, and on the state of standing flora in other patches $\{Y_{c',t}, c' \neq c\}$ (or a sub-set of these patches). This transition probability, $P(Z_{c,t} = z_{c,t} | Z_{c,t-1} = z_{c,t-1}, \boldsymbol{Y}_t^C = \boldsymbol{y}_t^C)$, is denoted as $A(z_{c,t-1}, z_{c,t}, \boldsymbol{y}_t^C)$, where $\boldsymbol{Y}_t^C = (Y_{1,t}, \ldots, Y_{C,t})$. The notation used in the MHMM-DF is summarized in Table 5.1.

Variable	Domain	Definition
C	$\mathbb{N}^*$	Number of chains (patches)
T	$\mathbb{N}^*$	Number of observation time steps
$Z_{c,t}$	$\Omega_Z = \{0, 1, \ldots \lvert\Omega_Z\rvert - 1\}$	Abundance class of the dormant population in patch c at time t
$Y_{c,t}$	$\Omega_Y = \{0, 1, \ldots \lvert\Omega_Y\rvert - 1\}$	Abundance class of the non-dormant population in patch c at time t
Z_t^C	Ω_Z^C	$\{Z_{c,t}, 1 \leq c \leq C\}$
Y_t^C	Ω_Y^C	$\{Y_{c,t}, 1 \leq c \leq C\}$

Table 5.1. *Notation of variables in the MHMM-DF*

Using this notation, the joint probability of the set of hidden and observed variables, from time $t = 0$ to time $t = T$, can be decomposed as follows:

$$\prod_{c=1}^{C} \gamma_0(z_{c,0}) \prod_{t=1}^{T} A(z_{c,t-1}, z_{c,t}, \boldsymbol{y}_t^C)\phi(z_{c,t-1}, y_{c,t})$$

5.2.3. *Parameterizing the model*

Generally, samples of standing flora are relatively limited in terms of the number of patches and/or years they cover due to the cost of sampling (in terms of time and

---

1 By convention, in what follows, the random variables will be represented by uppercase letters and the realizations of these variables will be shown in lowercase.

effort). For estimation purposes, we shall consider a parsimonious parametric version of the MHMM-DF.

First, let us consider the probability of emission $\phi(z_{c,t-1}, y_{c,t})$. For a fixed value of $z_{c,t-1}$, this will be modeled as a mixture of a Dirac on zero and a binomial distribution. If $z_{c,t-1} = 0$, then $y_{c,t-1} = 0$. If $z_{c,t-1} > 0$, then $y_{c,t-1}$ follows a binomial distribution of parameters $|\Omega_Y|$ and $p_{z_{c,t-1}}$. The probability p_z $(z \in \Omega_Z)$ itself is modeled by a logistic regression: $p_z = \frac{1}{1+\exp(-(\mu_0+\mu_1 z/|\Omega_Z|))}$, where μ_0 and μ_1 are two parameters. The choice of a binomial distribution was guided by the fact that only a small number of parameters are required. Furthermore, this distribution respects an increasing relationship between the abundance class of standing flora and that of the seed bank.

Similarly, for fixed values of $z_{c,t-1}$ and $\boldsymbol{y}_t^C$, the probability of transition A is modeled by a binomial distribution of parameters $|\Omega_Z|$ and $p_{z_{c,t-1}, \boldsymbol{y}_t^C}$. The second parameter is also modeled by a logistic regression. There are several possible options available for constructing this regression; only one will be presented here. First, there is a natural distinction to be made between the influence of $y_{c,t}$ and that of standing flora in patches other than c. Let $\boldsymbol{y}_t^{\bar{c}} = \{y_{c',t}, c' \neq c\}$ be the set of abundance classes of standing flora in patches other than c at time t. The regression model is a function of $z_{c,t-1}$, $y_{c,t}$ and $\boldsymbol{y}_t^{\bar{c}}$. It is possible to consider only a subset of $\boldsymbol{y}_t^{\bar{c}}$ as having an influence on $z_{c,t}$, for example, only the standing flora of patches which are geographically close to c. Next, as $\boldsymbol{y}_t^{\bar{c}}$ is a multidimensional variable, it can be summarized by a variable of dimension 1: $mean(\boldsymbol{y}_t^{\bar{c}})$, its mean. This is a measure of the mean capacity of colonization from other patches toward patch c. The logistic regression model is thus written as:

$$p_{z_{c,t-1}, \boldsymbol{y}_t^C} = \frac{1}{1+\exp(-(\nu_0 + \nu_1 \times \frac{z_{c,t-1}}{|\Omega_Z|} + \nu_2 \times \frac{y_{c,t}}{|\Omega_Y|} + \nu_3 \times \frac{mean(\boldsymbol{y}_t^{\bar{c}})}{|\Omega_Y|}))},$$

where ν_0, ν_1, ν_3 and ν_3 are four parameters.

Finally, the initial distribution γ_0 of the dormant state in patch c is modeled by a binomial distribution of parameters $|\Omega_Z|$ and p_τ, where

$$p_\tau = \frac{1}{1+\exp(-\tau)}.$$

This parameterization of the MHMM-DF uses seven parameters (see Table 5.2). Its generic identifiability can be demonstrated where $C > 2$ and $T > 7$ (see Le Coz *et al.* 2019).

5.2.4. *Linking the parameters of the model with the ecological parameters of the dynamics of an annual plant*

The interpretation of all seven parameters of the model is shown in Table 5.2. In addition to these parameters, other quantities of interest can be calculated with respect to this model, such as the probability of colonization of the local seed bank, p_{col} (whether by other patches or through exogenous colonization); the probability of the seed bank surviving between two time steps (s); and finally, the probability of germination (g).

Calculating p_{col}: the probability of colonization of the local seed bank, whether from other patches or exogenous colonization, is defined by

$$p_{col} = 1 - P(Z_{c,t} = 0 | Y_{c,t} = 0, Z_{c,t-1} = 0, mean(\boldsymbol{y}_t^{\bar{c}}) \in \Omega_Y). \qquad [5.1]$$

This is the probability of the seed bank being non-empty at time t, in the absence of seed bank at time $t - 1$ and in the absence of local standing flora.

Let p_{exo} be the probability of exogenous colonization of the seed bank, and $p_{neighbor}$ the probability of colonization by another patch close to c. A distinction can be made between these two types of colonization in the MHMM-DF:

$$p_{exo} = 1 - P(Z_{c,t} = 0 | Y_{c,t} = 0, Z_{c,t-1} = 0, mean(\boldsymbol{Y}_t^{\bar{c}}) = 0) \qquad [5.2]$$

and, supposing the exogenous colonization and neighbor colonization events to be independent,

$$\begin{aligned}(1 - p_{exo})(1 - p_{neighbor}) &= \\ P(Z_{c,t} = 0 | Y_{c,t} = 0, Z_{c,t-1} = 0, mean(\boldsymbol{Y}_t^{\bar{c}}) &\in \{1, ..., |\Omega_Y| - 1\}). \qquad [5.3]\end{aligned}$$

p_{col}, p_{exo} and $p_{neighbor}$ can thus be determined using equalities [5.1] to [5.3], as the right-hand side terms can be estimated by simulation.

Calculating s: considering exogenous colonization to be independent of the survival of the seed bank, we have

$$(1 - p_{exo})(1 - s) = P(Z_{c,t} = 0 | Y_{c,t} = 0, Z_{c,t-1} \in \Omega_Z, mean(\boldsymbol{Y}_t^{\bar{c}}) = 0)$$

Once p_{exo} has been calculated, we can then calculate s.

Calculating g: the probability g of dormant seeds germinating is defined as the probability of observing a non-zero abundance class of plants in the patch when the seed bank is not empty:

$$g = P(Y_{c,t} \neq 0 | Z_{c,t-1} \neq 0).$$

The different probabilities involved in calculating s and g can also be estimated by simulation.

Parameters	Interpretation for annual plants
Initial distribution $P(z_{c,0}) = \gamma_0(z_{c,0})$	
τ	Mean trend
Emergence from dormant state $P(y_{c,t}\|z_{c,t-1}) = \phi(z_{c,t-1}, y_{c,t})$	
μ_0	$1/(1+e^{-\mu_0-\mu_1/\|\Omega_Z\|})$ lower bound of the probability of germination
μ_1	Influence of the seed bank on the next standing population
Dynamics of the seed bank $P(z_{c,t}\|z_{c,t-1}, \boldsymbol{y}_t^C) = A(z_{c,t-1}, z_{c,t}, \boldsymbol{y}_t^C)$	
ν_0	Exogenous colonization by propagule rain
ν_1	Influence of the previous year's seed bank
ν_2	Influence of local seed production
ν_3	Influence of seed production in other patches

Table 5.2. *Interpretation of parameters of the MHMM-DF based on binomial distributions and logistic regression*

5.2.5. *Estimation*

The classic method for estimating the parameters of an HMM is via the EM (expectation–maximization, Dempster *et al.* 1977) algorithm. This iterative algorithm alternates between a step E, in which a conditional expectation is calculated for the current value of parameters, and a maximization step M, in which parameter values are updated. In the case of a hidden Markov chain, step E relies on the use of the forward–backward algorithm (Rabiner 1989), which uses the linear structure of the model alongside variable elimination principles to calculate the conditional probabilities of hidden variables, given observed values, in an efficient manner. Although the MHMM-DF may be seen as an HMM in which $\boldsymbol{Z}_t^C$ is the hidden variable at time t, the forward–backward algorithm is difficult to apply directly in this case due to the multi-dimensional nature of $\boldsymbol{Z}_t^C$. The size of the domain of $\boldsymbol{Z}_t^C$ means that implementation of step E rapidly becomes impossible.

However, the structure of the MHMM-DF is such that, conditional on observations, each hidden chain is independent. By rewriting of the equations in the forward–backward algorithm, step E can be considered as C independent forward–backward algorithms. This implementation remains exact, and its complexity is linear as a function of the number of patches (a direct application of the forward–backward to Z_t^C would result in exponential complexity).

Details of the formulas used in the EM algorithm for an MHMM-DF, along with a study of its behavior using simulated data, can be found in Le Coz *et al.* (2019).

5.2.6. *Model selection*

In practice, there are several choices that must be made when constructing an MHMM-DF for a given plant: whether or not the species is dormant, how far seeds can disperse, whether external factors affect dynamics and thus the parameters to estimate, etc. These choices can be made by estimating several variants of the model and selecting the version that maximizes a model selection criterion, such as the BIC (Schwarz 1978). All likelihood-based criteria are calculable, as this quantity is easy to evaluate using output from the forward–backward algorithm.

5.3. Dynamics of weed species in cultivated parcels

5.3.1. *Dormancy and weed management in agroecosystems*

Dormancy may be defined as an internal state of a seed that prevents it from germinating despite favorable thermal, gaseous and hydric conditions (Benech-Arnold *et al.* 2000). Dormancy is not a binary characteristic: a continuum exists between dormant and non-dormant states. An increasing level of dormancy corresponds to a reduction in the range of environmental conditions in which germination will occur, by a reduction in germination speed, and by a reduction in the percentage of seeds that will finally germinate. The form of dormancy by which a seed stock can be preserved is known as secondary dormancy[2], which is an effect of environmental conditions (Baskin and Baskin 1998). The intensity of dormancy is dependent on environmental conditions and differs between species. Generally speaking, high temperatures, short days, dry conditions and easy access to nutrients during seed setting by the mother plants results in low dormancy levels in the produced seeds (Baskin and Baskin 1998). Secondary dormancy is a means of synchronizing the germination phase with the cycle of the seasons. The classification of species by field germination time also reflects the periodicity of their dormancy. A distinction can be made between species for which the minimum period of dormancy ends in early fall (geraniums, blackgrass) or late fall (stickywilly, ivy-leaved speedwell), spring (curlytop knotweed) or summer (amaranths, crabgrass) or which

2 Not to be confused with primary dormancy, which occurs at the point where the seed reaches maturity on the mother plant, and often disappears in a matter of months.

have an almost non-existent dormant period (groundsel, common field-speedwell). As the majority of weed species have a physiological dormancy period, their level of dormancy is governed by seasonal variations in temperature and soil humidity. For example, for common knotgrass and pale persicaria, which both emerge in the spring, the cooler temperatures of fall mark the end of the fully dormant period and by mid-winter, the seed is no longer dormant. Germination will then occur as soon as soil temperature exceeds a certain threshold in the spring. The warm temperatures of late spring trigger the start of the dormant state, and germination capacity progressively diminishes through May and June, with maximum dormancy attained in early summer. Winter species (fall germination, spring flowering) have an opposite dormancy cycle, in which high temperatures signal the end of dormancy, and low temperatures during the winter trigger or maintain dormancy.

5.3.2. *Description of the data set*

Studies carried out in the experimental farm at Epoisses, France, from 2000 to 2017 (see Figure 5.2) focused on designing and testing low-herbicide cropping (without tillage, with or without mechanical weeding) and evaluating the agronomic, economic, environmental and social performances of these systems (Adeux *et al.* 2019). At the time of writing, this focus has shifted onto testing and evaluating pesticide-free systems via the CA-SYS platform (Cordeau *et al.* 2015a).

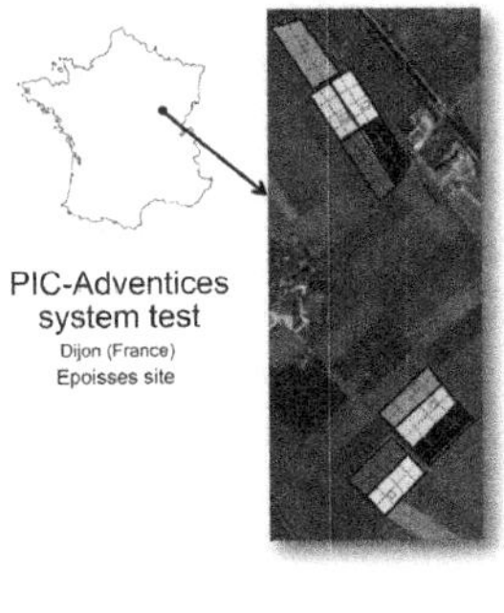

Figure 5.2. *Layout of the 10 fields and 90 patches in the experimental farm at Epoisses. The edges of each field are shown in black. Each field is broken down into nine patches (plus a tenth zone not used in this study). For a color version of this figure, see www.iste.co.uk/peyrard/ecology.zip*

For the purposes of our study, we shall use data concerning weed flora from 2000 to 2017. The experimental setup consisted of two groups of five fields. Each of these 10 fields was split into nine zones, corresponding to the patches in the MHMM-DF (see Figure 5.2). The density of weeds per square meter was measured each year, taken as the mean value of four quadrats, taken in different positions each time. Data were collected on a maximum of five occasions, before and after weeding. For a given year and plot, the data used to estimate the MHMM-DF are that relating to the final date of sampling on the plot. In 90% of cases, data were collected five times, and thus reflect the state of the weed species at the same point in the growing season. Pre-weeding observations account for less than 10% of the data.

Over the course of the whole experimental period (2000–2017), a total of 70 weed species were identified. For the purposes of this study, we have chosen to focus on seven species: *Chenopodium album*, *Solanum nigrum*, *Alopecurus myosuroides*, *Fallopia convolvulus*, *Aethusa cynapium*, *Galium aparine* and *Polygonum aviculare* (Cordeau *et al.* 2020). These species are diverse in terms of germination period and in the potential quantity of seeds produced by each individual plant. They were also among the most common species found in the test fields at Epoisses. Abundance classes were defined by splitting the interval of observed data values (log transformed) into four intervals of equal width (the first class is that of a null abundance). Thus, $|\Omega_Y| = 5$ and we have also chosen $|\Omega_Z| = 5$. The class boundaries for the seven weed species in our study are shown in Table 5.3. Other, less empirical divisions have been tested (Barralis scale, search for the closest stepwise function to the data histogram), but have proven unsatisfactory due to the potential existence of empty classes.

Class	1	2	3	4	5
Alopecurus myosuroides	0	$]0, 2.39]$	$]2.39, 15.59]$	$]15.59, 101.44]$	> 101.44
Chenopodium album	0	$]0, 1.16]$	$]1.16, 3.65]$	$]3.65, 11.48]$	> 11.48
Solanum nigrum	0	$]0, 1.36]$	$]1.36, 5.05]$	$]5.05, 18.71]$	> 18.71
Fallopia convulus	0	$]0, 1.47]$	$]1.47, 5.89]$	$]5.89, 23.59]$	> 23.59
Aethusa cynapium	0	$]0, 1.06]$	$]1.06, 3.03]$	$]3.03, 8.68]$	> 8.68
Galium aparine	0	$]0, 1.39]$	$]1.39, 5.22]$	$]5.22, 19.64]$	> 19.64
Polygonum aviculare	0	$]0, 1.19]$	$]1.19, 3.85]$	$]3.85, 12.44]$	$> 12, 44$

Table 5.3. *Boundaries of the five abundance classes for the seven weed species in our study*

A study of the Bravais–Pearson correlation between the abundance of observed flora in a patch at time t and that of other patches at time $t + 1$ shows that this value becomes negligible at distances longer than 60 m, whatever the species in question. Thus, in calculating the function $mean(\boldsymbol{y}_t^{\bar{c}})$, we only include patches at a distance of less than 60 m from patch c. Based on this choice, a patch may be colonized from a maximum of six other patches. The total number of patches is 88 (90, minus two patches in which no flora was observed in 2003).

5.3.3. *Comparison with an HMM with independent patches*

For illustrative purposes, let us consider how the inclusion of inter-patch colonization affects the estimation of the parameters of the dynamics of a weed species. To do this, we shall compare the estimated values of p_{col} and s obtained for the MHMM-DF and for the model presented in Pluntz *et al.* (2018). The latter is an HMM that operates at parcel level: patches are considered to be independent, and colonization occurs uniquely in the form of a propagule rain. This means that, unlike the MHMM-DF, this model is non-spatial and p_{col} cannot be broken down into p_{exo} and $p_{neighbor}$. Furthermore, this model uses a presence–absence representation for both the seed bank and standing flora, and the probability of setting seed is fixed at 1.

The estimators for both models are shown in Table 5.4. We see that the estimated colonization value is higher using MHMM-DF than with the non-spatial HMM for all seven species. The main reason for this lies in the fact that the probability of seed setting is set at 1 for all weed species in the non-spatial model, but in practice, certain species rarely set seed, or only set seed in certain environments. Thus, while *Chenopidium album* and *Solanum nigrum*, for example, can grow in wheat fields, they will not set seed in this habitat. In this case, the overestimation of the seed supply resulting from seed setting is offset "mechanically" by a lower estimation of the probability of colonization.

Species	Model	s	p_{col}	p_{nbor}	g
Alopecurus myosuroides	Non-spatial	0.56	0.25	-	0.61
	MHMM-DF	0.54	0.37	0.10	0.64
Chenopodium album	Non-spatial	0.79	0.15	-	0.33
	MHMM-DF	0.66	0.23	0.11	0.38
Solanum nigrum	Non-spatial	0.71	0.15	-	0.36
	MHMM-DF	0.72	0.49	0.22	0.21
Fallopia convulus	Non-spatial	0.82	0.19	-	0.61
	MHMM-DF	0.80	0.31	0.02	0.61
Aethusa cynapium	Non-spatial	0.76	0.11	-	0.55
	MHMM-DF	0.61	0.24	0.08	0.44
Galium aparine	Non-spatial	0.79	0.16	-	0.52
	MHMM-DF	0.77	0.47	0.26	0.48
Polygonum aviculare	Non-spatial	0.59	0.17	-	0.50
	MHMM-DF	0.55	0.23	0.15	0.44

Table 5.4. *Probabilities of survival, colonization and germination from a dormant state in the non-spatial HMM and MHMM-DF models*

Furthermore, the estimation of survival is lower in the MHMM-DF than in the non-spatial HMM for all seven species. Finally, we note that the estimated value for germination after dormancy is similar for both models; this is as expected, as this value is not dependent on local abundance.

5.3.4. *Influence of crops on weed dynamics*

Twenty different crop species were sown over the 17-year study period: 73.53% fall crops (oilseed rape, wheat, barley, triticale), 11.76% summer crops (soy beans, sorghum, sunflower), 11.76% spring crops (peas, barley) and 2.94% perennial crops (alfalfa). It is important to note that data for the same weed species in different crops were not collected at the same time of year (Adeux *et al.* 2019). Weed germination is highly dependent on environmental conditions, notably those resulting from tillage (Cordeau *et al.* 2017b). Some of the fields at Epoisses were not tilled, a factor known for limiting the germination of most weed crops (Cordeau *et al.* 2015b), particularly in cases where seed/soil contact is poor and in dry summers (Cordeau *et al.* 2018). Furthermore, Donohue (2005) found that seed germination is highly dependent on the season. Thus, the density of a weed species varies according to the planted crop (see Table 5.6). For this reason, we split crops into two groups for the purposes of our study: winter crops (sown in the fall) and summer crops (sown in spring-summer). It is thus possible, for each weed species, to distinguish between the dynamics of patches bearing a summer crop and patches with a winter crop. However, the type of crop grown in neighboring patches, contributing to the colonization of each local patch, will not be taken into account.

Our results show that the dynamics of all species except *Alopecurus myosuroides* (blackgrass) and *Aethusa cynapium* are best described by a model which takes account of the planted crop type (see Table 5.5). The exception of the species cited above is likely due to their generalist and widespread nature. Furthermore, the case of *Alopecurus myosuroides* is different from that of other weeds, as seeds were sown in a stock zone of every field at Epoisses, independent of the crop, at the beginning of the experiment.

Three of the seven species in our study exhibit stronger germination from dormancy in the presence of a summer crop (see Table 5.6): *Chenopodium album*, *Solanum nigrum* and *Polygonum aviculare*. The others (*Alopecurus myosuroides*, *Fallopia convulus*, *Aethusa cynapium* and *Galium aparine*) perform better in this sense in the presence of a winter crop. This corresponds to the fact that the first three species naturally emerge in spring-summer, while the four remaining species are essentially autumnal. *Fallopia convulus* is a spring plant, but emerges particularly early, and is frequently encountered in winter crops.

Finally, for *Chenopodium album* and *Polygonum aviculare*, the probability of the seed bank being maintained from year t to year $t + 1$ is significantly higher when the crop at $t + 1$ is a summer crop (see Table 5.6).

BIC	Model with crop seasonality	Model without crop seasonality
Alopecurus myosuroides	2697	**2634**
Chenopodium album	**1655**	1670
Solanum nigrum	**1778**	1875
Fallopia convolvulus	**3066**	3088
Aethusa cynapium	2804	**2795**
Galium aparine	**2090**	2095
Polygonum aviculare	**2296**	2301

Table 5.5. *BIC selection values for models with and without taking account of crop seasonality for each species. The best BIC value for each species is in bold*

Species	Crop season	s	g	Density
Alopecurus myosuroides	Winter	0.55	0.69	9.08
	Summer	0.46	0.51	7.66
Chenopodium album	Winter	0.47	0.12	0.86
	Summer	0.85	0.80	1.23
Solanum nigrum	Winter	0.37	0.20	1.36
	Summer	0.26	0.87	3.93
Fallopia convulus	Winter	0.74	0.63	4.17
	Summer	0.80	0.62	2.46
Aethusa cynapium	Winter	0.60	0.58	2.44
	Summer	0.67	0.32	1.36
Galium aparine	Winter	0.60	0.84	2.21
	Summer	0.74	0.21	4.18
Polygonum aviculare	Winter	0.40	0.37	2.02
	Summer	0.85	0.72	1.23

Table 5.6. *Probabilities of survival and emergence from a dormant state in the MHMM-DF model taking account of local crop seasonality. The final column shows plant density per square meter (averaged over all parcels for the same season)*

5.4. Discussion and conclusion

In this chapter, we presented a statistical model with hidden variables for the dynamics of annual plants, combining two approaches: an ecological approach, based on metapopulation dynamics which, in its initial form, ignores the state of the seed bank; and an agronomic approach, which takes account of the seed bank but generally models colonization as a propagule rain. The different processes involved (survival of the seed bank, emergence from a dormant state, colonization) are described in a relatively simple manner, involving the use of qualitative data (abundance classes). Nevertheless, this study demonstrates that an HMM-based approach can be used to extract relevant information about the dynamics of several species with marked biological differences (germination period, phenology and seed production).

We opted for a parsimonious approach to modeling, using a binomial distribution combined with logistic regression. The advantage of this approach is that the number of parameters remains constant, whatever the number of abundance classes. Other, more classic parameterizations of abundance classes, such as a cumulative logit distribution or a cumulative Beta distribution, could also be used (Herpigny and Gosselin 2015); however, more parameters would need to be estimated in these cases.

The fact that the effect of the set of abundance classes of standing flora in neighboring patches has been summarized here using the effect of the average class is another point for discussion. Several abundance class vectors may result in the same average value, without necessarily corresponding to the same colonization potential. Le Coz (2019) uses a different form of single-value representation, arranging the vector of abundance classes in alphabetical order. Each possible configuration of neighborhood states is thus distinct from the others, and configurations are arranged in order of increasing colonization potential. Another option in this case would be to introduce a weighting for abundance classes based on the distance between the colonizing and receiving parcels.

One advantage of the MHMM-DF lies in the combination of a clear representation of the complex spatiotemporal dependences present in the dynamics of plant populations including dormancy behaviors with an estimation method of reasonable complexity (linear with respect to C, the number of patches). Le Coz *et al.* (2019) present a study of the behavior of the EM algorithm using simulated data, and show that, except in cases where the hidden variables take only extreme values, the quality of the estimators produced is good. However, this method – like all spatiotemporal methods – only works if enough observable population data are available. From this perspective, the Epoisses dataset is exceptional in terms of the quantity and quality of data available. Note that it is possible to compensate for sampling limitations, in terms of the number of time steps, by increasing the spatial range of the sample, since the number of hidden chains is not a limiting factor in EM. The estimators obtained should be interpreted having in mind of the spatial scale of patches and of the distances between patches. The estimated probability of colonization will differ depending on the relation between patch surface and the average seed dispersal distance, which may be similar or significantly different. Finally, the quality of estimators is highly dependent on the quality of observations, that is, of the abundance class for standing flora in a patch. Different species have different occurrence frequencies. Increasing the sampled surface (e.g. to a patch size of 2,000 m) may limit the number of observations in which zero specimens of a species are identified, but this reduces the quality of the abundance estimation, even if abundance classes are used. Reducing patch size (e.g. using quadrants) improves the precision of abundance estimations, but also increases the number of zero observations.

Applying the MHMM-DF model to the estimation of weed species dynamics in crops provides us with a clearer understanding of these dynamics. In previous studies, colonization has been modeled as a propagule rain, with no spatial localization of the source(s) of these seeds. In agronomic terms, our results highlight the influence of crop choice on the dynamics of a weed species, as one might expect, since the chosen crop determines the sowing period and thus the weed species which are likely to germinate in the same patch (Cordeau *et al.* 2017b). Colonization also has a role to play; given the size and distribution of patches, this mostly occurs on an intraparcellar level or across plots in the area surrounding the boundary, which is entirely logical given the known dispersal distances of the species in question.

The R code used to analyze the data from the experimental farm at Epoisses is available to download[3]. This code includes observations, transformed into abundance classes, for each of the seven species and all 88 patches covered by our study, along with inter-patch spatial correlation data. The code can be used to estimate parameters for the MHMM-DF with or without crop seasonality and the subsequent ecological parameters p_{col}, p_{exo}, s and g. Finally, the code also includes calculations of the ecological parameters of the non-spatial HMM model (Pluntz *et al.* 2018).

Several packages in R include functions that can be used to estimate the parameters of a one-dimensional HMM. In theory, these functions could be used to estimate the MHMM-DF, but the algorithmic complexity of transforming this model into a one-dimensional HMM is exponential as a function of the number of chains; computation would thus be impossible for cases involving large numbers of patches. The package `ensembleFHMM` can be used for estimation in the case of a multi-dimensional HMM of the factorial type (Ghahramani and Jordan 1997), in which the hidden chains emit a shared observation for each time step. In this case, the chains become dependent, conditional on observations; this is not true of the MHMM-DF. The `CHMM` package in R is designed for another type of multi-dimensional HMM, coupled HMMs (Brand *et al.* 1997), in which hidden chains are correlated, conditionally or otherwise, with observations. Once again, the dependency structure is different to that used in an MHMM-DF, so this package cannot be used with our model. Nevertheless, the CHMM and FHMM packages – both created in the context of signal processing – may prove helpful in relation to other spatiotemporal dynamics in ecology.

In methodological terms, it is important to note that the MHMM-DF framework itself can be extended for use with other dynamics, not simply those of annual plants,

3 https://oliviergimenez.github.io/code_livre_variables_cachees/.

while continuing to provide an exact EM estimation of linear complexity with respect to the number of patches. The only condition is that the hidden chains are independent, conditional on observations. For example, assuming that the survival of the observable population is possible from one step to the next, the model can be used to study the dynamics of perennial plants or cryptic species. The conditional independence of hidden chains is also maintained if an additional form of colonization, from an observable population to the observable population of another patch, is added; this is particularly relevant for certain animal species, which go through a cryptic stage.

5.5. Acknowledgments

This work was partly funded by the ANR AGROBIOSE (ANR-2013-0001) and the Région Occitanie. The authors wish to thank Guillaume Adeux for synthesizing and supplying 17 years of data from the PIC-adventices experiment, collected by the UMR BGA team then by the UMR Agroécologie, Dijon (Dominique Meunier, Florence Strbik, Francois Dugué). The PIC-Adventices experiment was launched in 2000 by Nicolas Munier-Jolain (INRAE, UMR Agroécologie) and by the team at the experimental farm (Phillippe Chamoy, Pascal Farcy, Luc Biju-Duval, Benjamin Pouilly, Claude Sarrasin, Alain Berthier), with support from the ANR CoSAC (ANR-15-CE18-0007), the European Horizon 2020 research and innovation program (no. 727321, IWM PRAISE) and, more recently, by the French Investissement d'Avenir project (ISITE-BFC Agroecologie en BFC, ANR-15-IDEX-03).

5.6. References

Adeux, G., Munier-Jolain, N.M.D., Farcy, P., Carlesi, S., Barberi, P.S.C. (2019). Diversified grain-based cropping systems provide long-term weed control while limiting herbicide use and yield losses. *Agronomy for Sustainable Development*, 39(4), 42.

Baskin, C. and Baskin, J. (1998). *Seeds: Ecology, Biogeography, and Evolution of Dormancy and Germination*. Academic Press, San Diego.

Benech-Arnold, R., Sanchez, R., Forcella, F., Kruk, B., Ghersa, C. (2000). Environmental control of dormancy in weed seed banks in soil. *Field Crops Research*, 67, 105–122.

Borgy, B., Reboud, X., Peyrard, N., Sabbadin, R., Gaba, S. (2015). Dynamics of weeds in the soil seed bank: A hidden Markov model to estimate life history traits from standing plant. *PLoS ONE*, 10(10).

Brand, M., Oliver, N., Pentland, A. (1997). Coupled hidden Markov models for complex action recognition. In *Proceedings of IEEE Computer Society Conference on Computer Vision and Pattern Recognition*, IEEE, San Juan, PR, 994–999.

Cordeau, S., Deytieux, V., Lemanceau, P., Marget, P. (2015a). Towards the establishment of an experimental research unit on agroecology in France. *Aspects of Applied Biology*, 128, 271–273.

Cordeau, S., Guillemin, J., Reibel, C., Chauvel, B. (2015b). Weed species differ in their ability to emerge in no-till systems that include cover crops. *Annals of Applied Biology*, 166, 444–455.

Cordeau, S., Smith, R., Gallandt, E., Brown, B., Salon, P., DiTommaso, A., Ryan, M. (2017a). Disentangling the effects of tillage timing and weather on weed community assembly. *Agriculture*, 7, 66.

Cordeau, S., Smith, R., Gallandt, E., Brown, B., Salon, P., DiTommaso, A., Ryan, M. (2017b). Timing of tillage as a driver of weed communities. *Weed Science*, 65, 504–514.

Cordeau, S., Wayman, S., Reibel, C., Strbik, F., Chauvel, B., Guillemin, J. (2018). Effects of drought on weed emergence and growth vary with seed burial depth and presence of a cover crop. *Weed Biology and Management*, 18, 12–25.

Cordeau, S., Adeux, G., Meunier, D., Strbik, F., Dugué, F.B.H., Vieren, E., Louviot, G., Munier-Jolain, N. (2020). Weed density of 7 major weeds in the long-term integrated weed management cropping system experiment of Dijon-Epoisses (2000–2017). INRAE, Data Portal.

Dempster, A., Laird, N., Rubin, D. (1977). Maximum likelihood from incomplete data via the EM algorithm. *Journal of the Royal Statistical Society*, 39(1), 1–38.

Donohue, K. (2005). Seeds and seasons: Interpreting germination timing in the field. *Seed Science Research*, 15(3), 175–187.

Freckleton, R. and Watkinson, A. (2002). Large-scale spatial dynamics of plants: Metapopulation regional ensembles and patchy populations. *Journal of Ecology*, 90, 419–434.

Freville, H., Choquet, R., Pradel, R., Cheptou, P.-O. (2013). Inferring seed bank from hidden Markov models: New insights into metapopulation dynamics in plants. *Journal of Ecology*, 101(6), 1572–1580.

Gallandt, E. (2006). How can we target the weed seedbank? *Weed Science*, 54, 588–596.

Ghahramani, Z. and Jordan, M. (1997). Factorial hidden Markov models. *Machine Learning*, 29(2–3), 245–273.

Herpigny, B. and Gosselin, F. (2015). Analyzing plant cover class data quantitatively: Customized zero-inflated cumulative beta distributions show promising results. *Ecological Informatics*, 26, 18–26 [Online]. Available at: http://www.sciencedirect.com/science/article/pii/S1574954114001629.

Lamy, T., Gimenez, O., Pointier, J., Jarne, P., David, P. (2013). Metapopulation dynamics of species with cryptic life stages. *American Naturalist*, 181(4), 479–491.

Le Coz, S. (2019). Modélisation de la dynamique des adventices dans un agroécosystème. PhD Thesis, INRAE, Paris.

Le Coz, S., Cheptou, P.-O., Peyrard, N. (2019). A spatial Markovian framework for estimating regional and local dynamics of annual plants with dormant stage. *Theoretical Population Biology*, 127, 120–132.

Leibold, M. and Miller, T. (2004). From metapopulations to metacommunities. *Ecology, Genetics and Evolution of Metapopulation*, Hanski, I. and Gaggiotti, O.E., (eds), Academic Press, San Diego.

Levins, R., Vagaggini, D., Zarattini, P., Mura, G. (1969). Some demographic and genetic consequences of environmental heterogeneity for biological control. *Bulletin of the Entomological Society of America*, 15(3), 237–240.

Lewis, J. (1973). Longevity of crop and weed seeds: Survival after 20 years in soil. *Weed Research*, 13, 179–191.

Manna, F., Pradel, R., Choquet, R., Fréville, H., Cheptou, P.-O. (2017). Disentangling the role of seed bank and dispersal in plant metapopulation dynamics using patch occupancy surveys. *Ecology*, 98(10), 2662–2672.

Petit, S., Boursault, A., Le Guilloux, M., Munier-Jolain, N., Reboud, X. (2011). Weeds in agricultural landscapes. A review. *Agronomy for Sustainable Development*, 31, 309–317.

Pluntz, M., Coz, S.L., Peyrard, N., Pradel, R., Choquet, R., Cheptou, P.-O. (2018). A general method for estimating seed dormancy and colonisation in annual plants from the observation of existing flora. *Ecology Letters*, 21(9), 1311–1318.

Rabiner, L. (1989). A tutorial on hidden Markov models and selected applications in speech recognition. *Proceedings of the IEEE*, 77, 257–286.

Schwarz, G. (1978). Estimating the dimension of a model. *The Annals of Statistics*, 6, 461–464.

Vegis, A. (1964). Dormancy in higher plants. *Annual Review of Plant Physiology*, 15, 185–224.

6

Using Latent Block Models to Detect Structure in Ecological Networks

Julie AUBERT[1], Pierre BARBILLON[1], Sophie DONNET[1] and Vincent MIELE[2]

[1] *Paris-Saclay University, AgroParisTech, INRAE, UMR MIA-Paris, France*

[2] *Laboratoire de Biométrie et Biologie Évolutive, Université Lyon 1, CNRS, UMR5558, Villeurbanne, France*

6.1. Introduction

The study of inter-species relations within an ecosystem has been an important theme in ecology since the publication of Charles Elton's seminal work on food chains (Elton 1927). Relationships between species can be represented as a network of interactions, with nodes taking the place of biological entities (generally species) of interest and edges (or links) representing the interactions in question. The analysis of ecological networks is the focus of much recent ecological literature and has enjoyed an upsurge in interest in the last 20 years, notably through the work of Dunne *et al.* (2002). Networks of interactions may be studied at microscopic level (e.g. microbial network) or at higher levels (e.g. plant–pollinator relationships). They may represent a wide variety of interaction types, such as mutualism, competition or predation. Figure 6.1 shows a food web made up of 106 different animal and plant species observed on the coast of Chile. Networks may be unipartite, representing interactions within a single group of given species (as in the food web shown in Figure 6.1), or bipartite, if they describe relations between two different types of entities (e.g. plant–pollinator networks, host–parasite networks, etc.). In what follows, all networks will be considered unipartite unless otherwise stated.

Analyzing the structure of a network may be crucial for understanding the organization of an ecosystem, notably via the extraction of its organizational structure in summary form. Two types of approach may be used in this context (Miele *et al.* 2019). The first family of approaches are descriptive, based on different metrics used to characterize properties at node level (e.g. centrality) or across the whole network (e.g. nestedness (Almeida-Neto *et al.* 2008) or modularity (Barber 2007)). In contrast, the second group consists of aggregative approaches, focusing on "zooming out" the network in order to detect groups of species with similar interaction properties. While it is possible to detect compartments (or modules, Krause *et al.* (2003)), several recent works have followed the pathway laid down by Allesina and Pascual (2009) based on the notion of "functional" groups of species sharing the same connection pattern (in the sense of ecological interactions). For example, species at the same trophic level may be seen as a group. In this context, a distinction may be made between purely algorithmic approaches and those based on generative probabilistic models. The latter category is the subject of this chapter. More specifically, we shall focus on a mixture-based probabilistic approach using stochastic block models (SBMs). Introduced in the field of sociology by Snijders and Nowicki (1997), SBMs work on the premise that nodes are divided into latent blocks (groups, clusters, etc.) containing entities with similar connection profiles. SBMs have been shown to be effective for identifying groups of nodes or blocks, which play the same role in a network, irrespective of the type of structure in question. SBMs have been extended for bipartite networks in the form of latent block models (LBM) (Govaert and Nadif 2003) or bipartite SBM (biSBM; Larremore *et al.* (2014)), which is equivalent (Wyse *et al.* 2017). Blocks are introduced as latent variables in both models. Unlike metric-based approaches, this probabilistic approach is agnostic, in that it does not aim to highlight a particular structure, but rather to group nodes that behave in the same way in the network, irrespective of the behavior itself. While this approach was slow to be adopted in the context of ecological networks, it has recently gained in popularity and is at the heart of an increasing number of new applications (Allesina and Pascual 2009; Baskerville *et al.* 2011; Kéfi *et al.* 2016; Michalska-Smith *et al.* 2018; Ohlmann *et al.* 2019; Ohlsson and Eklöf 2020; Miele *et al.* 2020).

In what follows, we shall present SBMs for both unipartite and bipartite binary networks and illustrate their flexibility (i.e. their capacity to describe a wide variety of structures). We shall then discuss the statistical inference of these models: in this way, starting with an observed network, a description of the relations between groups of nodes can be obtained along with the groups themselves. Our approach will be illustrated using two case studies of ecological networks: the Chilean food web mentioned above (Kéfi *et al.* 2016), which is a directed network, and a bipartite plant–pollinator network based on the observed interactions between plants and pollinating insects in Tenerife in 2012 (Carstensen *et al.* 2018).

6.2. Formalism

Ecological networks are made up of nodes, representing biological entities of interest (generally species) and edges representing the interaction being studied. Unipartite networks (interactions with a single given group of species) may be directed, if the relationship being represented is oriented (as in the case of a food chain), or non-directed, for example if the relationship is mutualist. In bipartite networks, interactions occur between two predetermined groups of species (plants–pollinators, hosts–parasites, etc.). In this case, the connection between two nodes is often asymmetric, for example, "pollinates" or "is pollinated by", and there are no edges between nodes of the same nature.

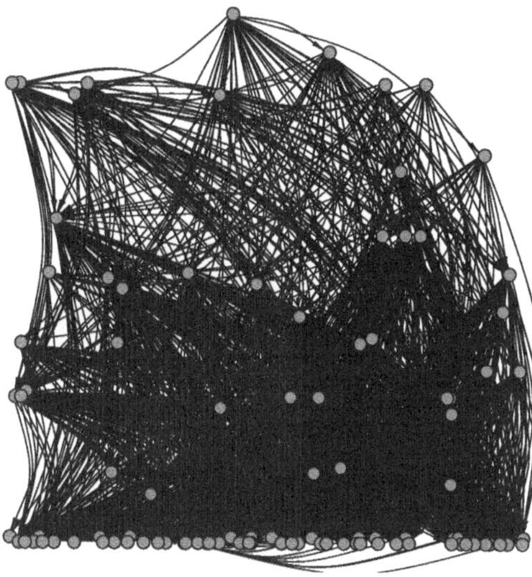

Figure 6.1. *The Chilean network: 1,362 trophic interactions observed in the inter-tidal zone of the Pacific coast of Chile, involving 106 sessile or mobile animal and plant species (Kéfi* et al. *2016). For a color version of this figure, see www.iste.co.uk/peyrard/ecology.zip*

Furthermore, interactions may be treated as binary data (observed/non-observed) or count data (e.g. as a number of observed interactions). For the purposes of this study, we shall focus on binary interactions; the model can be extended to weighted networks (corresponding to count data) with no great difficulty, and this case will be discussed later. Finally, note that nodes or edges may be described by a set of covariates.

A network may be encoded as a matrix $\mathbf{y}$ such that $y_{ij} \neq 0$ if the entity i is in interaction with entity j, or 0 otherwise. In a unipartite network, the species in the rows are the same as those in the columns, and the matrix (square) is known as an *adjacency matrix*. Generally, in this case, $y_{ii} = 0, \forall i$. If the relation is oriented, then $y_{ij} \neq y_{ji}$, otherwise $\forall(i,j), y_{ij} = y_{ji}$ and $\mathbf{y}$ is symmetrical. In a bipartite network, the nodes (entities) in the rows are not the same as those in the columns, and $\mathbf{y}$ is a rectangular matrix known as an *incidence matrix* or bi-adjacency matrix.

The encoding of a network as a matrix is illustrated in Figures 6.2 and 6.3 for a non-directed unipartite network and a bipartite network, respectively.

$$y = \begin{pmatrix} 0\,1\,0\,1 \\ 1\,0\,1\,1 \\ 0\,1\,0\,0 \\ 1\,1\,0\,0 \end{pmatrix}$$

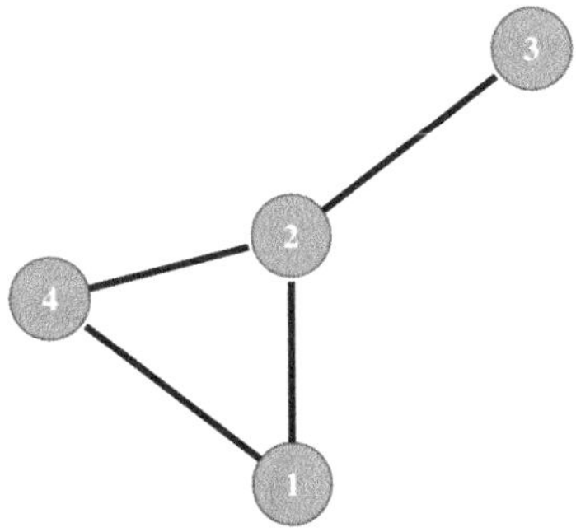

Figure 6.2. *Adjacency matrix and corresponding representation of the non-directed binary network. For a color version of this figure, see www.iste.co.uk/peyrard/ecology.zip*

$$y = \begin{pmatrix} 1\,1\,1\,1\,0 \\ 0\,0\,1\,1\,1 \\ 0\,0\,0\,0\,1 \end{pmatrix}$$

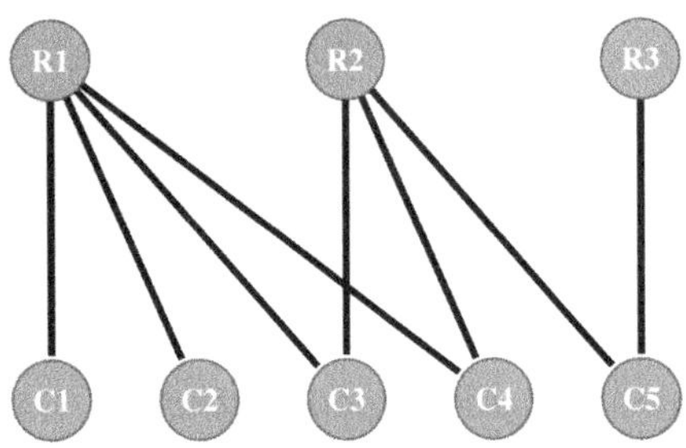

Figure 6.3. *Incidence matrix and corresponding representation of the bipartite binary network. For a color version of this figure, see www.iste.co.uk/peyrard/ecology.zip*

6.3. Probabilistic mixture models for networks

In this section, we shall present SBMs for both network types. These models work on the assumption that the matrix $\mathbf{y}$ that describes the network is a realization of a random variable $\mathbf{Y}$, of which the distribution is given.

6.3.1. *SBMs for unipartite networks*

Consider a unipartite network of size n. Let $i = 1, \ldots, n$ be the nodes and let $Y_{ij} \in \{0, 1\}$ be the random variable characterizing the relationship between the pair of nodes i and j. In an SBM, nodes are assumed to be divided into K latent, that is, non-observed, groups (blocks). A variable Z_i is associated with each node $i = 1, \ldots, n$ such that $Z_i = k$ if node i belongs to group k.

Following the classic approach for mixture models, the values of Z_i are assumed to be independent and identically distributed such that $\forall i = 1, \ldots, n, \forall k = 1, \ldots, K$

$$P(Z_i = k) = \pi_k \qquad [6.1]$$

where $\pi_k \in [0, 1]$ are such that $\sum_{k=1}^{K} \pi_k = 1$. The parameter π_k thus corresponds to the probability of belonging to group k. Conditional to these latent variables, the SBM considers the interactions Y_{ij} to be independent random variables such that:

$$P(Y_{ij} = 1 | Z_i = k, Z_j = k') = \gamma_{kk'} \qquad [6.2]$$

Thus, the probability of interaction between any pair of nodes belonging to blocks k and k', respectively, is $\gamma_{kk'}$.

REMARK.– If the relationship of interest is non-oriented, then the model is defined only for $i < j$ since $Y_{ji} = Y_{ij}$ for all (i, j) and we have $\gamma_{kk'} = \gamma_{k'k}$ for all (k, k'). In the opposite case, equation [6.2] is defined for all pairs (i, j) such that $i \neq j$ and $\gamma_{kk'}$ may be different from $\gamma_{k'k}$.

In what follows, we shall use the notation $\boldsymbol{\gamma} = (\gamma_{kk'})_{k,k'=1,\ldots,K}$ and $\boldsymbol{\pi} = (\pi_k)_{k=1,\ldots,K}$. Let $\boldsymbol{\theta} = (\boldsymbol{\gamma}, \boldsymbol{\pi})$ be the parameters of the model and $\mathbf{Z} = (Z_i,)_{i=1\ldots,n}$ the latent variables representing block memberships.

ILLUSTRATION.– Figures 6.4 and 6.5 show realizations of the SBM for different parameter values (the parameters are shown on the left, the matrix on the right). For the network shown in Figure 6.4, matrix $\boldsymbol{\gamma}$ contains a diagonal that is stronger than the rest of the matrix, giving a marked modular tendency. The network shown in Figure 6.5 is typical of a food web structure, and the blocks are essentially equivalent to different levels of the food chain. These particular structures were chosen as both are commonly encountered in ecology. The variety of possible choices for parameter $\boldsymbol{\gamma}$ implies that the SBM is highly flexible. Furthermore, the fact that the model is generative makes it easy to test hypotheses by simulation.

$$\gamma = \begin{pmatrix} 0.5 & \varepsilon & \varepsilon \\ \varepsilon & 0.6 & \varepsilon \\ \varepsilon & \varepsilon & 0.7 \end{pmatrix}$$

$$\varepsilon = 0.1 \quad \pi = \left(\frac{1}{3}, \frac{1}{2}, \frac{1}{6}\right)$$

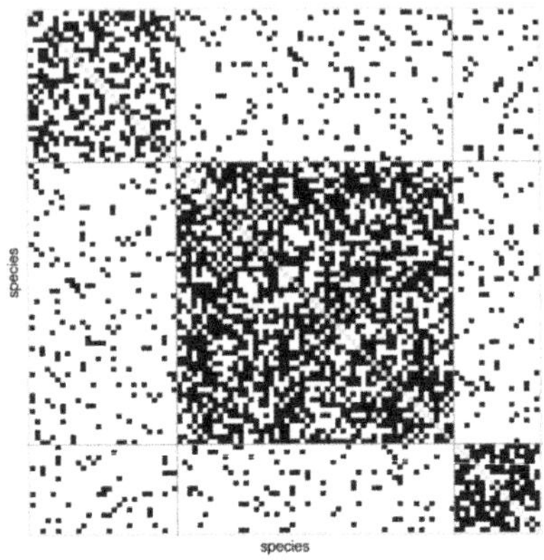

Figure 6.4. *Simulation of a modular network for the parameters shown on the left. A network realization is shown on the right in the form of an adjacency matrix ($y_{ij} = 0$ corresponds to a white square and $y_{ij} = 1$ to a black square)*

$$\gamma = \begin{pmatrix} \varepsilon & 0.5 & \varepsilon \\ \varepsilon & \varepsilon & 0.5 \\ \varepsilon & \varepsilon & \varepsilon \end{pmatrix}$$

$$\varepsilon = 0.05$$

$$\pi = \left(\frac{1}{3}, \frac{1}{3}, \frac{1}{3}\right)$$

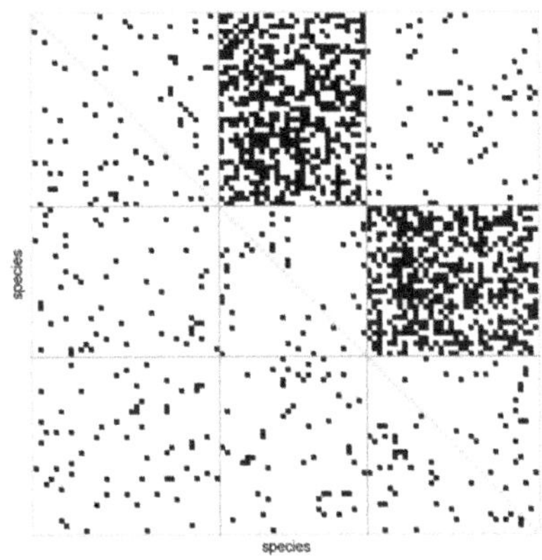

Figure 6.5. *Simulation of a food web for the parameters shown on the left. A realization of the network is shown on the right in the form of an adjacency matrix ($y_{ij} = 0$ corresponds to a white square and $y_{ij} = 1$ to a black square)*

6.3.2. *Stochastic block model for bipartite networks*

Now, let us consider the case of a bipartite network of size $n_L \times n_C$, with n_L nodes of a first type and n_C nodes of a second type. The model shown above can be rewritten asymmetrically, with the introduction of two sets of latent variables, representing a co-clustering in rows and columns. For all $i = 1, \ldots, n_L$, let Z_i be the variable such that $Z_i = k$ if the entity in row i belongs to group k. Similarly, for all $j = 1, \ldots, n_C$,

$W_j = g$ if j belongs to group g. (Z_i) and (W_j) are assumed to be independent and identically distributed such that for all $(i, k) \in \{1, \ldots, n_L\} \times \{1, \ldots, K\}$ and for all $(j, g) \in \{1, \ldots, n_C\} \times \{1, \ldots, G\}$,

$$P(Z_i = k) = \pi_k^L; \quad P(W_j = g) = \pi_g^C, \qquad [6.3]$$

where $\pi_k^L \in [0, 1]$ with $\sum_{k=1}^K \pi_k^L = 1$, and $\pi_g^C \in [0, 1]$ with $\sum_{g=1}^G \pi_g^C = 1$. The distribution of interactions is then defined conditionally on these groups: for all $(i, j) \in \{1, \ldots, n_L\} \times \{1, \ldots, n_C\}$,

$$P(Y_{ij} = 1 | Z_i = k, W_j = g) = \gamma_{kg}. \qquad [6.4]$$

Just as before, this model is sufficiently flexible to adapt to a wide range of topologies. For example, Figure 6.6 shows a nested bipartite network, in which specialist species are in interaction with generalist species. This nested structure is widespread in ecology. The simulated network shown in Figure 6.7 is more unusual, featuring a structure which is both nested and modular.

$$\gamma = \begin{pmatrix} 0.8 & 0.7 & 0.6 \\ 0.8 & 0.5 & \varepsilon \\ 0.6 & \varepsilon & \varepsilon \end{pmatrix}$$

$$\varepsilon = 0.03$$

$$\pi^L = \left(\frac{1}{3}, \frac{1}{3}, \frac{1}{3}\right)$$

$$\pi^C = \left(\frac{1}{3}, \frac{1}{2}, \frac{1}{6}\right)$$

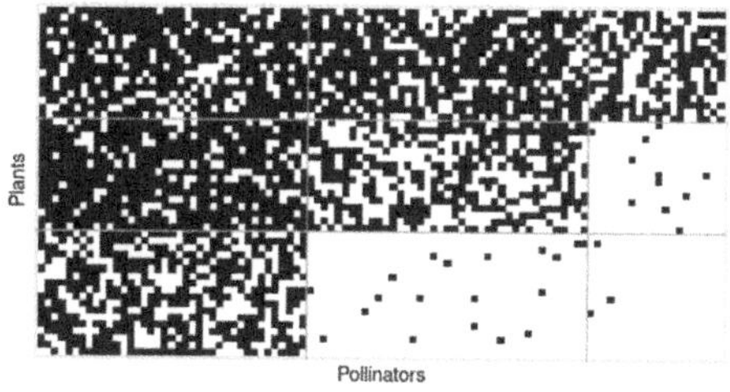

Figure 6.6. *Simulation of a nested bipartite network for the parameters shown on the left. A realization of the network is shown on the right in the form of an adjacency network ($y_{ij} = 0$ corresponds to a white square, $y_{ij} = 1$ to a black square)*

In this case, $\boldsymbol{\theta} = (\boldsymbol{\gamma}, \boldsymbol{\pi}^L, \boldsymbol{\pi}^C)$ is the set of parameters of the model $\mathbf{Z} = (Z_i,)_{i=1\ldots,n_L}$, and $\mathbf{W} = (W_j,)_{j=1\ldots,n_C}$ are the latent variables.

These two models are identified up to label switching.

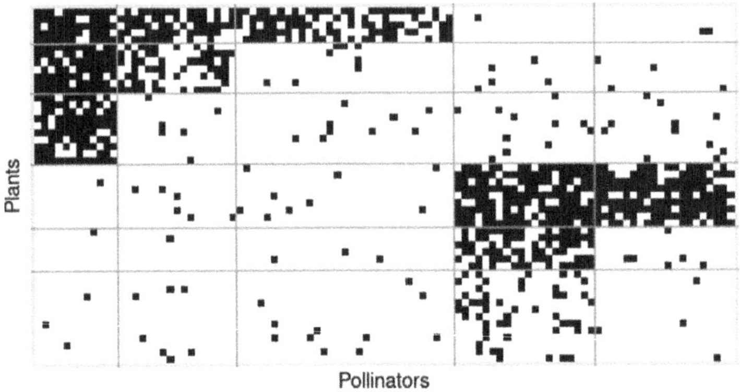

$$\gamma = \begin{pmatrix} 0.8 & 0.7 & 0.6 & \varepsilon & \varepsilon \\ 0.8 & 0.5 & \varepsilon & \varepsilon & \varepsilon \\ 0.8 & \varepsilon & \varepsilon & \varepsilon & \varepsilon \\ \varepsilon & \varepsilon & \varepsilon & 0.8 & 0.8 \\ \varepsilon & \varepsilon & \varepsilon & 0.6 & \varepsilon \\ \varepsilon & \varepsilon & \varepsilon & 0.2 & \varepsilon \end{pmatrix} \qquad \begin{aligned} \varepsilon &= 0.03 \\ \pi^L &= \left(\frac{1}{6}, \frac{1}{6}, \frac{1}{6}, \frac{1}{6}, \frac{1}{6}, \frac{1}{6}\right) \\ \pi^C &= \left(\frac{2}{15}, \frac{1}{5}, \frac{4}{15}, \frac{1}{5}, \frac{1}{5}\right) \end{aligned}$$

Figure 6.7. *Simulation of a bipartite network with a modular and nested structure for the parameters shown below. A realization of the network is shown at the top of the figure in the form of an adjacency network ($y_{ij} = 0$ corresponds to a white square, $y_{ij} = 1$ to a black square)*

6.4. Statistical inference

The aim, based on observations of the network $\mathbf{y}$, is to infer parameters and identify the latent variables. This is done by means of two processes. First, for a given number of blocks (K in the unipartite case, (K, G) in the bipartite case), we must identify the parameters that maximize likelihood, and the most probable groups given the observations $\mathbf{y}$; second, we must identify the optimal number of blocks, which involves a compromise between fitting and sparsity (i.e. the number of parameters in the model).

For simplicity's sake, the SBM parameter estimation process will only be presented for unipartite networks here. For an in-depth presentation of the bipartite case, see Govaert and Nadif (2008).

6.4.1. *Estimation of parameters and clustering*

First, let us consider a case with a fixed number of groups K. We wish to estimate parameters by maximizing the likelihood. The joint distribution of the observed variables $(Y_{ij})_{i,j=1,\ldots,n}$ and latent variables $\mathbf{Z}$ is easy to write because, conditional on the groups, the values of Y_{ij} are independent and follow a Bernoulli distribution (equation [6.2]), while variables $\mathbf{Z}$ are assumed to be independent (equation [6.1]):

$$\log p(\mathbf{y}, \mathbf{Z}; \boldsymbol{\theta}) = \log p(\mathbf{y}|\mathbf{Z}; \boldsymbol{\gamma}) + \log p(\mathbf{Z}; \boldsymbol{\pi})$$

with

$$\begin{aligned}\log p(\mathbf{y}|\mathbf{Z}; \boldsymbol{\theta}) &= \log \prod_{i=1}^{n} \prod_{j=1, j\neq i}^{n} \gamma_{Z_i Z_j}^{y_{ij}} (1 - \gamma_{Z_i Z_j})^{1-y_{ij}} \\ &= \sum_{i=1}^{n} \sum_{j=1, j\neq i}^{n} \sum_{k,k'=1}^{K} \mathbb{1}_{Z_i=k} \mathbb{1}_{Z_j=k'} (y_{ij} \log \gamma_{kk'} + (1 - y_{ij}) \\ &\quad \times \log(1 - \gamma_{kk'})), \\ \log p(\mathbf{Z}; \boldsymbol{\theta}) &= \sum_{i=1}^{n} \sum_{k=1}^{K} \mathbb{1}_{Z_i=k} \log \pi_k\end{aligned}$$

As the variables $\mathbf{Z}$ are non-observed, the likelihood of observations is obtained by integrating the complete likelihood $p(\mathbf{y}, \mathbf{Z}; \theta)$ against the set of possible values of the latent variables:

$$\ell(\mathbf{y}; \boldsymbol{\theta}) = \sum_{\mathbf{Z} \in \{1,\ldots,K\}^n} \exp\{\log p(\mathbf{y}, \mathbf{Z}; \boldsymbol{\theta})\} \qquad [6.5]$$

This integration over latent variables [6.5] implies that the calculation and, to an even greater extent, the maximization of this function are complex from a numerical perspective (the sum contains K^n terms). In this case, the expectation–maximization (EM) algorithm is particularly suitable for likelihood maximization. However, in step E of the algorithm, the distribution of the latent variables $\mathbf{Z}$ must be calculated conditional on the observations $\mathbf{y}$. In the case of the SBM, this distribution is not factorized (conditional on the observations, the latent variables Z_i are not independent), and this prevents direct application of the algorithm. The variational version of the EM algorithm offers an effective solution to this problem (Daudin *et al.* 2008). In this case, the conditional distribution $p(\mathbf{Z}|\mathbf{y}; \boldsymbol{\theta})$ is approximated within a family of simpler distributions where Z_i are assumed independent, and a lower bound of the likelihood is maximized [6.5]. Like any EM algorithm, the variational EM (VEM) is highly sensitive to initialization. It is, therefore, important to test multiple, non-random initializations. Other inference methods have been described in the literature (see Lee and Wilkinson 2019).

The outputs from the VEM algorithm are estimated parameters $\widehat{\theta}$ and, for each node, a probability of belonging to each block. Nodes can then be clustered into groups using the maximum *a posteriori* rule, which consists of placing each observation into the group to which it is most likely to belong:

$$\widehat{Z}_i = \underset{Z_i=1,\dots,K}{\arg\max}\ \widehat{p}(\mathbb{1}_{Z_i=k}|\mathbf{y};\widehat{\boldsymbol{\theta}})$$

We note $\widehat{\mathbf{Z}} = (\widehat{Z}_i)_{i=1,\dots,n}$. Bickel *et al.* (2013) have demonstrated the consistency of variational estimators in the SBM, whereas Mariadassou *et al.* (2010) extended the method to weighted networks. A presentation of inference methods in bipartite networks can be found in Govaert and Nadif (2008).

6.4.2. *Model selection*

The number of blocks K in the SBM (or the numbers of blocks K and G in the case of bipartite networks) is selected using a penalized likelihood criterion. Classic criteria, such as AIC or BIC, aim to establish a compromise between model fitting and complexity. In the case of unsupervised classification, the integrated classification likelihood (ICL) criterion (Biernacki *et al.* 2000) presents the advantage of aiming to optimize classification quality in addition to fitting data to the model, while applying a penalty for model size. Furthermore, this criterion uses complete likelihood (which is easy to calculate) rather than the likelihood of observations, which relies on integration with respect to the latent variables. For a directed network, this criterion is written as:

$$\mathrm{ICL}(K) = \log p(\mathbf{y}, \widehat{\mathbf{Z}}; \widehat{\boldsymbol{\theta}}) - \frac{1}{2}\left\{(K-1)\log n + K^2 \log\left(n^2 - n\right)\right\}$$

The term $(K-1)\log n$ in the penalty corresponds to the classification, and thus to parameters $(\pi_1, \dots, \pi_K)$ with a sum equal to 1. Term $K^2 \log\left(n^2 - n\right)$ corresponds to connections, where K^2 is the size of the matrix $(\gamma_{kk'})_{k,k'}$ and $n^2 - n$ is the number of observations $(\{i, j = 1, \dots, n; i \neq j\})$. These parameter numbers must be adapted to the case of a non-directed network in order to account for the symmetry of the matrices.

For a bipartite network, this criterion is written as:

$$\text{ICL}(K, G) = \log p(y, \widehat{\mathbf{Z}}, \widehat{\mathbf{W}}; \widehat{\theta}) \\ -\frac{1}{2}\{(K-1)\log n_L + (G-1)\log n_C + KG\log(n_L n_C)\}$$

The term $(K-1)\log n_L + (G-1)\log n_C$ corresponds to the two classifications, while the quantity $KG\log(n_L n_C)$ corresponds to interactions, and thus takes account of the size of the matrix $(n_L \times n_C)$.

This criterion has been widely used in the context of random network models (Daudin *et al.* 2008; Mariadassou *et al.* 2010; Keribin *et al.* 2015), proving its efficiency. Other criteria have been proposed in the literature (Hu *et al.* 2019). The inference of the SBM (i.e. parameter estimation, clustering for a fixed number of blocks and selection of a number of blocks using the ICL criterion) has been implemented in R using the sbm package (Chiquet *et al.* 2020). This package is an extension of the blockmodels package (Leger 2014) for simple and biSBM inference.

6.5. Application

In this section, we shall apply the SBM analytical method described above to two ecological networks: a food web and a plant–pollinator network.

6.5.1. *Food web*

Our case study covers all 1,362 trophic interactions observed in the intertidal zone on the Pacific coast of Chile, between 106 different, sessile or mobile, animal or plant species (Kéfi *et al.* 2016). The model selection procedure resulted in an SBM made up of seven blocks of differing sizes. By definition, these blocks group together species with similar connection patterns; we also see that the ecological characteristics of species within a block are similar.

The first block (B1) corresponds to "superpredators" (the top of the food chain), which have no predators except for rare predation connections between members of the block itself (shown as a loop in Figure 6.8). Nevertheless, there is a high level of taxonomic variation within the block, which features species as different from one another as anemones and gulls: taxonomy (or phylogeny) is not taken into account in the SBM approach to species grouping. Blocks B2 and B3 contain sessile mollusc species, which are targets of superpredators and consume algae (see below). The difference between the two blocks, as identified by the SBM, lies in the degree of generalism: species in block B2 consume a wider variety of prey, including those in B6, which are not consumed by species in block B3. The species in block B4 are also sessile, including small crustaceans, such as barnacles, and mollusks, such as mussels, which filter water to feed on plankton and are preyed upon by a variety of predator species. Block B5 is mostly made up of crabs, which are the exclusive targets of certain superpredators such as birds, as we see from the single connection leading to block B5 in Figure 6.8 (block B4, on the other hand, is the destination point of multiple connections). Blocks B6 and B7 contain basal algae species, including brown and red algae, respectively; as we have seen, these provide food for various mollusk species.

The SBM model thus offers a means of summarizing the complexity inherent in the observation of over a thousand interactions. By interpreting the parameters of the model (the probabilities of interactions between each pair of blocks), a synthetic description of the ecosystem can be created; Figure 6.8 is easier to read than Figure 6.1. This description is then used alongside additional information, such as taxonomy and ecological traits.

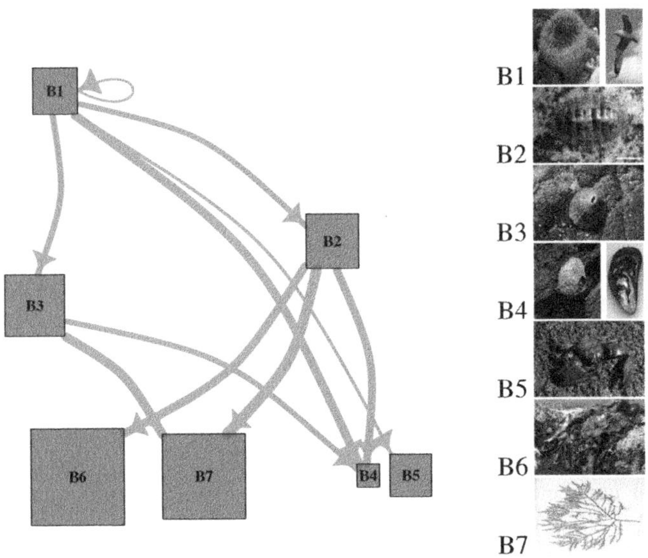

Figure 6.8. *Schematic representation (based on Picard* et al. *2009) of the estimated seven-block SBM for the Chilean network (Kéfi* et al. *2016). Left: Each square is a block, and the thickness of the lines represents the probability of interactions between each pair of blocks (above a threshold of 0.1, for the sake of clarity). Right: Types of species typical of those found in each block. From top to bottom: Anemone and gull (B1), chiton (B2), Fissurella (B3), Balanus and mussel (B4), crab (B5), Laminariale (B6) and red algae (B7). For a color version of this figure, see www.iste.co.uk/peyrard/ecology.zip*

6.5.2. *A bipartite plant–pollinator network*

In this case, the object of study is a network of interactions between plants and pollinating insects, observed in Tenerife in 2012 (Carstensen *et al.* 2018). In this study, interactions are binarized (i.e. the actual number of visits is not taken into account), but the same approach may be used in a way which takes account of the strength of interactions (the results in this case may be different). As the network is bipartite, we have chosen to use a biSBM or LBM, using the ICL criterion; applying this criterion gives a four-class model with two plant BLOCKS and two pollinator BLOCKS.

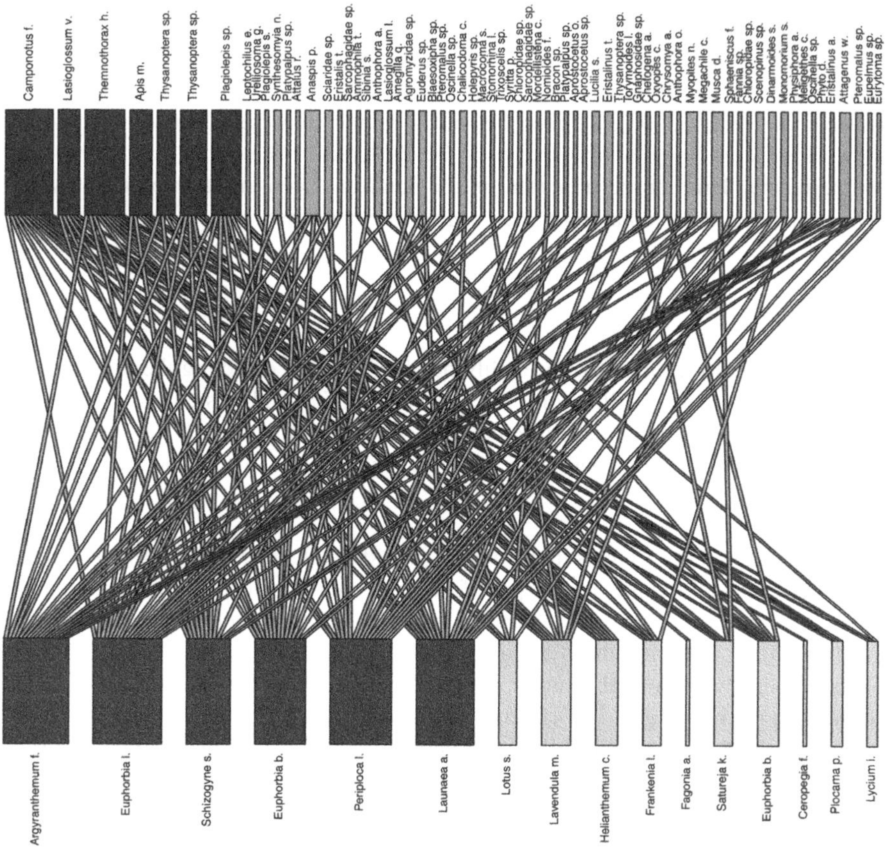

Figure 6.9. *Classic representation obtained using the R* `bipartite` *package. Different colors are used to show species belonging to the core (dark red, dark green) and to the periphery, as described by a four-block biSBM model. Pollinators are shown at the top, with plants at the bottom. For a color version of this figure, see www.iste.co.uk/peyrard/ecology.zip*

The proposed block structure adopts a core–periphery structure, which is particularly common in plant–pollinator interaction networks (Miele *et al.* 2020). The first block consists of a small number of insect species (Figure 6.9, in red), which account for the majority of edges with all plants. Unsurprisingly, the European honey bee forms part of this BLOCK. Similarly, one block of plants contains species involved in a large number of interactions with the majority of insects (Figure 6.9, dark green), such as the daisy *Argyranthemum frutescens*. These two blocks form the "core" of the network. The second BLOCK of insects is made up of "peripheral" species connected to the core BLOCK of plants (Figure 6.9, orange), while peripheral plants are linked to the core BLOCK of insect species (Figure 6.9, light green).

The LBM model thus describes a core–periphery architecture, well known in ecology, which is traditionally observed in terms of a measure of nestedness (Bascompte *et al.* 2003). It goes beyond a simple representation of the overall network structure, as it allows species to be clustered as belonging to the core or to the periphery. This can be important in the context of conservation policies, where species may be targeted based on their degree of generalism.

6.6. Conclusion

Stochastic block models offer highly flexible tools for detecting the macroscopic structure of a network. Structures such as modularity and nestedness are clearly observed in cases where such structures are traditionally expected; nevertheless, as no assumptions are made concerning network organization, other structures may also be revealed.

The use of latent variables in this model provides a flexible framework, which can easily be extended to take other types of network data into account. The distribution of the edges can notably be adapted for use with weighted networks (Mariadassou *et al.* 2010) by replacing the Bernoulli distribution in equations [6.2] and [6.4] by a Poisson, negative binomial (Aubert *et al.* 2021) or Gaussian distribution, according to the type of the weight. The effect of covariates can also be taken into account by adopting a kind of generalized linear model. For example, consider that each pair of nodes (i, j) is associated with a set of covariates in the vector $\mathbf{x}_{ij}$. These covariates can be integrated into the model as follows:

$$P(Y_{ij} = 1 | Z_i = k, Z_j = k') = \phi(\gamma_{kk'} + \mathbf{x}_{ij}^T \beta)$$

In this case, the clustering represents the variability of the connection phenomenon beyond that which can be explained by the covariates. Writing the model in this way also permits correction by node degree. Using a covariate to integrate a node-specific effect, the clustering process will detect any residual structure above and beyond the variability resulting from the number of connections (Karrer and Newman 2011). Another possibility is to consider that latent variables have an effect on covariates as well as on the connections in the network. In this case, covariates are taken into account in the inference process in order to improve clustering performance (Binkiewicz *et al.* 2017).

Multi-layer networks (Pilosof *et al.* 2017) involve the joint observation of multiple networks of interactions. These networks may be used to represent several different types of interactions between the same species (same nodes); in this case, we speak of a multiplex network. The SBM approach has been extended for use with multiplex networks by Kéfi *et al.* (2016) and Barbillon *et al.* (2016). Another type of multi-layer network is the multipartite network: these are the natural generalization of bipartite networks. In this case, the data used are obtained from joint observation of multiple networks representing the interactions between species belonging to pre-defined groups. These networks themselves may be bipartite between two predefined groups, or unipartite, for interactions within a single group. Dáttilo *et al.* (2016) has analyzed a dataset of this type, covering interactions between plants and pollinators, plants and ants, and plants and seed dispersing birds. This type of analysis is possible because of the extension of SBM models to multipartite networks (Bar-Hen *et al.* 2020); the model is also sufficiently general to handle a variety of other configurations. Predefined groups may be partitioned into sub-groups (blocks), taking account of the role played by different species in all of the networks in which they participate. In cases where interactions are observed over a period of time or a range of different locations, the resulting networks are said to be dynamic or spatialized, respectively. SBMs have been extended for both temporal and spatial cases (Matias and Miele 2017; Kim *et al.* 2018; Longepierre *et al.* 2019).

Another advantage of probabilistic models is the ease with which they can be integrated into more detailed models. The network observation process can be modeled and linked to the SBM generative model, meaning that SBM models can be inferred even in cases where data are missing. Furthermore, this approach may remove sampling bias if the observation process is missing not at random; for this, the observation process itself must be modeled (Tabouy *et al.* 2020). Once the SBM has been inferred, non-observed values can then be predicted. This approach is particularly helpful in the context of multi-layer networks where layers are observed with varying levels of precision. One benefit of joint inference is that the best-observed layers can be used to improve the overall inference of the model, improving prediction quality for the least well-observed layers.

Finally, note that SBMs are a special case of probabilistic network models including latent variables. They differ from other alternatives in that latent variables are considered to be discrete, corresponding to structural equivalence groups. Other latent variable models have been developed (Matias and Robin 2014); one notable example is the latent space model (Hoff *et al.* 2002), in which the latent variables associated with the nodes are considered to exist in a continuous space, and the probabilities of connection between nodes depend on the proximity between their associated latent variables.

6.7. References

Allesina, S. and Pascual, M. (2009). Food web models: A plea for groups. *Ecology Letters*, 12(7), 652–662.

Almeida-Neto, M., Guimaraes, P., Guimaraes Jr, P.R., Loyola, R.D., Ulrich, W. (2008). A consistent metric for nestedness analysis in ecological systems: Reconciling concept and measurement. *Oikos*, 117(8), 1227–1239.

Aubert, J., Schbath, S., Robin, S. (2021). Model-based biclustering for overdispersed count data with application in microbial ecology. *Methods in Ecology and Evolution*, 12(6), 1050–1061.

Bar-Hen, A., Barbillon, P., Donnet, S. (2020). Block models for generalized multipartite networks: Applications in ecology and ethnobiology. *Statistical Modelling* [Online]. Available at: https://doi.org/10.1177/1471082X20963254.

Barber, M.J. (2007). Modularity and community detection in bipartite networks. *Physical Review E*, 76(6), 066102.

Barbillon, P., Donnet, S., Lazega, E., Bar-Hen, A. (2016). Stochastic block models for multiplex networks: An application to a multilevel network of researchers. *Journal of the Royal Statistical Society*, 180, 295–314.

Bascompte, J., Jordano, P., Melián, C.J., Olesen, J.M. (2003). The nested assembly of plant–animal mutualistic networks. *Proceedings of the National Academy of Sciences*, 100(16), 9383–9387.

Baskerville, E.B., Dobson, A.P., Bedford, T., Allesina, S., Anderson, T.M., Pascual, M. (2011). Spatial guilds in the Serengeti food web revealed by a Bayesian group model. *PLoS Comput Biol*, 7(12), e1002321.

Bickel, P., Choi, D., Chang, X., Zhang, H. (2013). Asymptotic normality of maximum likelihood and its variational approximation for stochastic blockmodels. *The Annals of Statistics*, 41(4), 1922–1943.

Biernacki, C., Celeux, G., Govaert, G. (2000). Assessing a mixture model for clustering with the integrated completed likelihood. *IEEE Transactions on Pattern Analysis and Machine Intelligence*, 22(7), 719–725.

Binkiewicz, N., Vogelstein, J.T., Rohe, K. (2017). Covariate-assisted spectral clustering. *Biometrika*, 104(2), 361–377.

Carstensen, D.W., Trøjelsgaard, K., Ollerton, J., Morellato, L.P.C. (2018). Local and regional specialization in plant–pollinator networks. *Oikos*, 127(4), 531–537.

Chiquet, J., Donnet, S., Barbillon, P. (2020). SBM: Stochastic Blockmodels. R package version 0.2.2 [Online]. Available at: https://CRAN.R-project.org/package=sbm.

Dáttilo, W., Lara-Rodríguez, N., Jordano, P., Guimarães, P.R., Thompson, J.N., Marquis, R.J., Medeiros, L.P., Ortiz-Pulido, R., Marcos-García, M.A., Rico-Gray, V. (2016). Unravelling Darwin's entangled bank: Architecture and robustness of mutualistic networks with multiple interaction types. *Proceedings of the Royal Society of London B: Biological Sciences*, 283(1843) [Online]. Available at: https://royalsocietypublishing.org/doi/10.1098/rspb.2016.1564.

Daudin, J.J., Picard, F., Robin, S. (2008). A mixture model for random graphs. *Statistics and Computing*, 18(2), 173–183.

Dunne, J.A., Williams, R.J., Martinez, N.D. (2002). Food-web structure and network theory: The role of connectance and size. *Proceedings of the National Academy of Sciences*, 99(20), 12917–12922.

Elton, C. (1927). *Animal Ecology*. Sidgwick and Jackson, London.

Govaert, G. and Nadif, M. (2003). Clustering with block mixture models. *Pattern Recognition*, 36(2), 463–473.

Govaert, G. and Nadif, M. (2008). Block clustering with Bernoulli mixture models: Comparison of different approaches. *Computational Statistics and Data Analysis*, 52(6), 3233–3245.

Hoff, P.D., Raftery, A.E., Handcock, M.S. (2002). Latent space approaches to social network analysis. *Journal of the American Statistical Association*, 97(460), 1090–1098.

Hu, J., Qin, H., Yan, T., Zhao, Y. (2019). Corrected Bayesian information criterion for stochastic block models. *Journal of the American Statistical Association*, 115(532), 1771–1783.

Karrer, B. and Newman, M.E. (2011). Stochastic blockmodels and community structure in networks. *Physical Review E*, 83(1), 016107.

Kéfi, S., Miele, V., Wieters, E.A., Navarrete, S.A., Berlow, E.L. (2016). How structured is the entangled bank? The surprisingly simple organization of multiplex ecological networks leads to increased persistence and resilience. *PLoS Biology*, 14(8), 1–21.

Keribin, C., Brault, V., Celeux, G., Govaert, G. (2015). Estimation and selection for the latent block model on categorical data. *Statistics and Computing*, 25(6), 1201–1216.

Kim, B., Lee, K.H., Xue, L., Niu, X. (2018). A review of dynamic network models with latent variables. *Statistics Surveys*, 12, 105–135.

Krause, A.E., Frank, K.A., Mason, D.M., Ulanowicz, R.E., Taylor, W.W. (2003). Compartments revealed in food-web structure. *Nature*, 426(6964), 282–285.

Larremore, D.B., Clauset, A., Jacobs, A.Z. (2014). Efficiently inferring community structure in bipartite networks. *Physical Review E*, 90(1), 012805.

Lee, C. and Wilkinson, D.J. (2019). A review of stochastic block models and extensions for graph clustering. *Applied Network Science*, 4(1), 122.

Leger, J.-B. (2014). Wmixnet: Software for clustering the nodes of binary and valued graphs using the stochastic block model. Preprint arXiv:1402.3410.

Longepierre, L. and Matias, C. (2019). Consistency of the maximum likelihood and variational estimators in a dynamic stochastic block model. *Electronic Journal of Statistics*, 13(2), 4157–4223.

Mariadassou, M., Robin, S., Vacher, C. (2010). Uncovering latent structure in valued graphs: A variational approach. *The Annals of Applied Statistics*, 4(2), 715–742.

Matias, C. and Miele, V. (2017). Statistical clustering of temporal networks through a dynamic stochastic block model. *Journal of the Royal Statistical Society: Series B (Statistical Methodology)*, 79(4), 1119–1141.

Matias, C. and Robin, S. (2014). Modeling heterogeneity in random graphs through latent space models: A selective review. *ESAIM: Proceedings and Surveys*, 47, 55–74.

Michalska-Smith, M.J., Sander, E.L., Pascual, M., Allesina, S. (2018). Understanding the role of parasites in food webs using the group model. *Journal of Animal Ecology*, 87(3), 790–800.

Miele, V., Matias, C., Robin, S., Dray, S. (2019). Nine quick tips for analyzing network data. *PLoS Computational Biology*, 15(12), e1007434.

Miele, V., Ramos-Jiliberto, R., Vázquez, D.P. (2020). Core–periphery dynamics in a plant–pollinator network. *Journal of Animal Ecology*, 89(7), 1670–1677.

Ohlmann, M., Miele, V., Dray, S., Chalmandrier, L., O'Connor, L., Thuiller, W. (2019). Diversity indices for ecological networks: A unifying framework using Hill numbers. *Ecology Letters*, 22(4), 737–747.

Ohlsson, M. and Eklöf, A. (2020). Spatial resolution and location impact group structure in a marine food web. *Ecology Letters*, 23(10), 1451–1459.

Picard, F., Miele, V., Daudin, J.-J., Cottret, L., Robin, S. (2009). Deciphering the connectivity structure of biological networks using MixNet. *BMC Bioinformatics*, 10(6), S17.

Pilosof, S., Porter, M.A., Pascual, M., Kéfi, S. (2017). The multilayer nature of ecological networks. *Nature Ecology & Evolution*, 1(4), 0101.

Snijders, T.A.B. and Nowicki, K. (1997). Estimation and prediction for stochastic blockmodels for graphs with latent block structure. *Journal of Classification*, 14(1), 75–100.

Tabouy, T., Barbillon, P., Chiquet, J. (2020). Variational inference for stochastic block models from sampled data. *Journal of the American Statistical Association*, 115(529), 455–466.

Wyse, J., Friel, N., Latouche, P. (2017). Inferring structure in bipartite networks using the latent blockmodel and exact ICL. *Network Science*, 5(1), 45–69.

7

Latent Factor Models: A Tool for Dimension Reduction in Joint Species Distribution Models

Daria BYSTROVA[1], Giovanni POGGIATO[1,2], Julyan ARBEL[1] and Wilfried THUILLER[2]

[1] *University of Grenoble Alpes, Inria, CNRS, Grenoble INP, Laboratoire Jean Kuntzmann (LJK), France*

[2] *University of Grenoble Alpes, CNRS, University of Savoie Mont Blanc, Laboratoire d'Ecologie Alpine (LECA), France*

7.1. Introduction

Understanding how species are distributed in space has been one of the main goals of ecology. In particular, investigating which factors drive species distributions within communities, across regions or along environmental gradients can improve our understanding of fundamental ecological processes underlying such patterns, as well as our ability to anticipate future biodiversity changes (Thuiller *et al.* 2013; Guisan *et al.* 2017). When building models to explain and predict the distribution of organisms, we necessarily need to ask the same questions as the early biogeographers. It is now clear that three main conditions need to be met for a species to occupy a site and maintain populations (see Figure 7.1, Pulliam 2000; Lortie *et al.* 2004; Soberon 2007):

– the species has to physically reach the site, that is, to access the region (Barve *et al.* 2011);

Statistical Models for Hidden Variables in Ecology,
coordinated by Nathalie PEYRARD and Olivier GIMENEZ.

– the abiotic environmental conditions (i.e. temperature, precipitation, and so on) must be physiologically suitable for the species;

– the biotic environment (interactions with other species) must be suitable for the species.

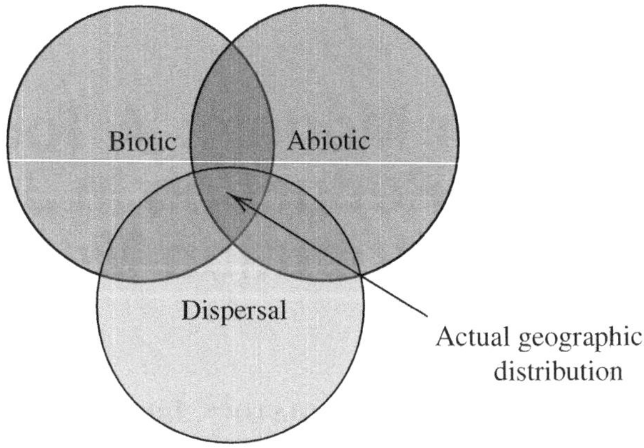

Figure 7.1. *The three factors that determine the actual distribution of a species (Soberon and Peterson 2005). For a color version of this figure, see www.iste.co.uk/peyrard/ecology.zip*

The first condition is a matter of species *dispersal* capacity from those areas previously occupied by the species. It includes the biogeographic history of the species, and thus all factors limiting its distribution from the place where it first originated, such as barriers to migration, biotic and abiotic dispersal vectors, rare long distance dispersal, etc.

The second condition is the matter of abiotic *habitat suitability* for the target species, which means that the combination of abiotic environmental variables at the site – often referred to as environmental suitability – are within the range of environmental conditions that the species requires to grow and maintain viable populations. These suitable environmental conditions are what ecologists call the *environmental niche* (Hutchinson 1957).

The third condition concerns *biotic interactions*, that is, interactions with other organisms, either positive (commensalism, mutualism) or negative (competition, predation), which themselves are influenced by the environment through their influence on all organisms in the local community.

From a statistical point of view, the most common tools to model how species are distributed in space are species distribution models (SDMs; Guisan and Thuiller

2005). There are a variety of SDMs that differ in their underlying statistical algorithms and flexibility (Guisan and Thuiller 2005; Merow *et al.* 2014; Guisan *et al.* 2017), but they all relate the presence or abundance, and sometimes the absence, of a species to a set of environmental variables and project this relationship in space and/or time. While SDMs have proven to be very useful and reliable in many different areas and fields (for reviews, see Guisan *et al.* 2017; Yates *et al.* 2018), they also have well-known limitations and assumptions that run counter to ecological niche theory (Guisan and Zimmermann 2000) and that may question the robustness of their predictions. A first major criticism of SDMs is that they model species independently of each other, making the assumption that species respond individualistically to the environment. As a result, SDMs can only capture the implicit combined effect of both abiotic and biotic environments. Despite these limitations, researchers have also used SDMs to predict species communities in space and time. In that case, single species predictions are simply stacked together (e.g. stacked SDMs; see Guisan and Thuiller 2005) by summing either the species' probabilities of occurrence (Calabrese *et al.* 2014) or the binary-transformed predictions (Guisan and Rahbek 2011). In the end, going from single species predictions to species communities commonly relies on a two-step procedure without any consideration of error propagation and without a joint-estimation of all model parameters.

With the increasing availability of community data (thanks to new sampling techniques like environmental DNA (eDNA) metabarcoding; see Taberlet *et al.* 2012), researchers now aim to model community as whole, and not as the stacked response of species (Clark *et al.* 2014). The species are then modeled together, giving birth to joint species distribution models (JSDMs; Pollock *et al.* 2014; Ovaskainen *et al.* 2017; Clark *et al.* 2017). These models estimate the relationship of each species with respect to environmental covariates through a regression, like SDMs, and additionally infer a correlation matrix among species from the regression residuals. This correlation matrix reflects species co-occurrence patterns not explained by the environmental predictors and may arise from model mis-specifications, missing covariates or, importantly, species interactions. Since the number of parameters in the residual correlation matrix scale quadratically with the number of species, these methods are computationally challenging. Latent factor models, which provide a low-rank approximation of this matrix, have naturally raised as a computationally efficient solution for JSDMs (Warton *et al.* 2015). In this chapter, we present latent factor models in the context of JSDMs, emphasizing their usefulness in community ecology. We apply latent factor models to plant species along 18 elevation gradients in the French Alps, belonging to the long-term observatory ORCHAMP (www.orchamp.osug.fr).

Within this book, two other chapters, Chapters 8 and 9, also develop methodologies for JSDMs. Chapter 8 focuses on the multivariate Poisson log-normal (PLN) model with abundance data, while ours essentially covers presence–absence data. Inference for this PLN model is done in a classical (non-Bayesian) setting with a variational approximation, while we follow a Bayesian approach and use a Markov chain Monte Carlo algorithm to sample from the posterior distribution, thus offering posterior credible intervals. Chapter 9 has a slightly different focus on how to combine predictors into components in order to lead to optimal learning. A classical (non-Bayesian) approach is used, and the case study tackles abundance data.

7.2. Joint species distribution models

To study species distribution, we relate a response variable $\mathbf{Y}_n$ to a set of p environmental covariates $\mathbf{X}_n = (X_{n\ell})_{\ell=1}^{p}$, at each site $n = 1, \ldots, N$. $\mathbf{Y}_n \in \mathbb{R}^S$ is a vector where each element Y_{ns} contains the observation for species $s = 1, \ldots, S$ at site n. Most JSDMs are based an extension of generalized linear models (GLMs), where they assume that the response variable is distributed as F, whose mean is given by a regression term and a residual multivariate random effect. For species s at site n, this is written:

$$Y_{ns} \sim F(\mu_{ns}, \phi_s) \quad [7.1]$$

$$g(\mu_{ns}) = \beta_{0s} + t(\boldsymbol{\beta}_s)\mathbf{X}_n + e_{ns} \quad [7.2]$$

$$\mathbf{e}_n \overset{\text{iid}}{\sim} \mathcal{N}_S(0, \boldsymbol{\Sigma}), \quad [7.3]$$

where F is the assumed distribution for the data with mean μ and dispersion parameter ϕ_s (which is usually not accounted for when modeling presence–absence data), and $t(\beta)$ denotes the transposition of β. Function g is called the *link function*. The vectors β_{0s} and $\boldsymbol{\beta}_s$ represent the intercept and regression coefficients for species s that describe the relationship between each species and the environmental covariates. Because of these coefficients, we can therefore define the suitable environmental conditions for each species, the *environmental niche*. Note here that the environmental covariates could also integrate the abundance or presence–absence of species (REF). Residual correlations among species are captured by Σ, a symmetric and positive-definite variance–covariance matrix (that has the constrain to be a correlation matrix for presence–absence data). The elements of $\boldsymbol{\Sigma}$ reflect species co-occurrence patterns that are not explained by the environmental predictors, and can arise from noise in the data, model misspecification, missing predictors and species associations.

The choice of the distribution F and the link function g depends on the response variable $\mathbf{Y}_n$ to be modeled. JSDMs typically model presence–absence, counts, biomass and many others due to the heterogeneity of ecological data and the sampling campaigns. For presence–absence data, most models assume a Bernoulli distribution and a probit link function (McCullagh and Nelder 1989, see GLM). However, this is quite common to replace the probit link function by a latent variable parameterization (Chib and Greenberg 1998) to make the model computationally more efficient. Since species community data may contain observations of species documented in multifarious ways (e.g. presence–absence and counts), several JSDMs have been implemented to address this challenge (Ovaskainen *et al.* 2017; Clark *et al.* 2017).

Interestingly, many JSDMs can model the regression coefficients hierarchically:

$$\boldsymbol{\beta}_s \overset{\text{iid}}{\sim} \mathcal{N}_p(\boldsymbol{\mu}, \boldsymbol{V}) \tag{7.4}$$

This allows for a share of information across species on their response to the environment, so that the estimation of the niche of rarely observed species could "borrow strength" from those of common species assuming that they do not behave fundamentally differently (Ovaskainen and Soininen 2011). Moreover, it is possible to account for functional traits and/or phylogeny by including them in $\boldsymbol{\mu}$ and/or $\boldsymbol{V}$ (e.g. see Chapter 6 of Ovaskainen and Abrego 2020 for detailed description).

7.3. Dimension reduction with latent factors

The model described above suffers from the "curse of dimensionality", since the covariance matrix scales quadratically with the number of species. Indeed, the number of free parameters of the covariance matrix when modeling S species is $S(S+1)/2$. For example, for $S = 100$ species, the number of parameters of the covariance matrix is more than 5,000. Nowadays, dealing with large datasets that contain observational data over space and time, the number of modeled species can easily exceed several thousands, making inference challenging and endless computational times. Hence, there is a need for dimension reduction approaches in JSDMs.

To address this challenge, several authors proposed a low-rank approximation of the covariance matrix of JSDMs through the use of latent factors (Warton *et al.* 2015). Starting from the original model [7.2], we assume a factorized representation of the residual random effects $\mathbf{e}_{ns}$, as a product of factor loadings and latent factors:

$$\mathbf{e}_{ns} = t(\mathbf{T}_s)\mathbf{Z}_n \quad \text{where} \quad \mathbf{Z}_n \overset{\text{iid}}{\sim} \mathcal{N}_K(0, \mathbf{I}_K) \tag{7.5}$$

The vector $\mathbf{T}_s \in \mathbb{R}^K$ is called the *factor loading* of species s; the collection of $t(\mathbf{T}_s)$, $s = 1, \ldots, S$, constitutes the rows of the so-called *factor loadings* matrix $\mathbf{T}$ (of dimension $S \times K$). The Gaussian random vectors $\mathbf{Z}_n \in \mathbb{R}^K$ are called *latent factors*. Crucially, note that under this factorized representation of the residual random effect, the residual covariance becomes now: $\mathbf{\Sigma} = \mathbf{T}t(\mathbf{T})$. By taking the number of latent factors $K \ll S$, the parameters to be modeled are drastically reduced.

Latent factors $\mathbf{Z}_n$ can be seen as a set of unmeasured covariates at site n, and the factor loadings $\mathbf{T}_s$ as the response of species s to these unmeasured covariates. A common (or opposite) response to these unmeasured covariates introduces a positive (or negative) correlation between species.

A critical feature of this dimension reduction is to appropriately select the number of latent factors. On the one hand, we need $K \ll S$ to reduce model complexity. On the other hand, we have to provide to the model the flexibility (that increases with a higher K) that is necessary to fully capture the required information from the data.

The number of factors controls the complexity of the model. The challenge is to find the appropriate number of factors such that the model is simple and tractable, yet appropriately capturing the covariance structure. Interestingly, this question arises also in most multivariate analyses where an optimal number of components has to be chosen. There are several approaches to address this issue in a Bayesian framework. One way is to initially fix K and then run a model selection with a range of K values. This is typically done by using information-theoretic criteria such as the deviance information criterion (DIC; Spiegelhalter *et al.* 2002) or the Watanabe–Akaike information criteria (WAIC; Watanabe and Opper 2010).

7.4. Inference

These models could be fitted either in the maximum likelihood framework or in the Bayesian one. The key difference between the two approaches is that maximum likelihood methods consider the model parameters as fixed (but unknown) quantities, while in the Bayesian approach they are considered as random (Ellison 2004). As a result, the Bayesian framework allows the introduction of a prior information on the parameters that might come from expert knowledge or previous studies. Bayesian methods also differ in the quantification of uncertainty: while maximum likelihood methods usually provide point parameter estimates and confidence interval, the Bayesian approach can provide the full distribution of the estimated parameters (the so-called posterior distribution).

Bayesian inference is particularly suitable in ecology due to its flexibility and computational tractability when dealing with highly complex models. Indeed, modeling nature is challenging due to the complexity and stochasticity of its underlying processes. This motivates the use of the Bayesian framework to analyze ecological data (Clark and Gelfand 2006). Introducing prior information in Bayesian models allows to incorporate various historical/external information and expert opinion for improving the models. Additionally, parameter estimations in these complex models are uncertain, and the Bayesian approaches are particularly suited for dealing with such an uncertainty.

As mentioned just above, a Bayesian framework implies to select suitable priors for model parameters: $\boldsymbol{\beta}_s, \boldsymbol{\Sigma}$. Incorporating prior information in the model could improve parameter estimates, but if priors are specified incorrectly, they could potentially bias the model, especially when only few observations are available. In practice, it is quite often difficult to specify correctly prior distributions reflecting prior knowledge. In this chapter, we present the case of more widespread or non-informative priors, but informative choices are also possible (Clark *et al.* 2017; Bystrova *et al.* 2021). The prior distribution for regression coefficients is usually a multivariate normal and an inverse Wishart for the covariance matrix, and all hyperpriors are chosen to be vague.

7.5. Ecological interpretation of latent factors

We described latent factors from the mathematical point of view, but what do they imply in term of ecological hypotheses and interpretation? In model [7.3], we described the residuals $\mathbf{e}_i$ for site i as a Gaussian vector whose covariance matrix $\boldsymbol{\Sigma}$ was unconstrained. This correlation reflects species co-occurrence patterns that are not explained by the environmental predictors, and may arise from model mis-specifications, missing covariates or species associations. We can also leverage on the non-independence between species to improve the co-occurrence and conditional predictions (see section 7.6). Latent factors not only allow to reduce the dimension of the model and to deal with a larger number of species, but they also yield crucial ecological insights.

First of all, this new representation still makes it possible to infer the residual covariance matrix among taxa: as shown previously, latent factors $\mathbf{T}$ factorize the covariance matrix into $\boldsymbol{\Sigma} = \mathbf{T}t(\mathbf{T})$. Therefore, species that are highly correlated have similar latent loadings. How can these latent loadings be interpreted?

In the latent factor representation of JSDMs, it is natural to think as the term $t(\mathbf{T}_s)\mathbf{Z}_n$ in equation [7.5] as a random effect term of a vector of latent covariates $\mathbf{Z}_n$ and their related species-specific coefficients $\mathbf{T}_s$. These latent covariates can be seen as *missing environmental predictors* and therefore provide a means of solving the longstanding problem of missing covariates modeling. In doing so, species with similar latent loadings share the same response to missing covariate and are thus expected to share similar occurrence patterns. Therefore, they are more correlated.

Latent factors can also be thought as *ordination axes* that represent the main axes of (co)variations of abundances across taxa. By forcing the number of latent factors to $K = 2$, it is possible to visualize on a biplot both the sites ordination, due to the latent variables $\mathbf{Z}_n$, and the ordination of taxa, with the latent loadings $\mathbf{T}_s$. Therefore, species that have close latent loadings will be close in the low dimensional space represented by the biplot, and therefore highly correlated. By evaluating this model-based ordination before and after the inclusion of measured environmental covariates, we can understand how much the co-occurrence suggested by an unconstrained model (i.e. without environmental covariates) can be explained by a shared response to environmental covariates.

7.6. On the interpretation of JSDMs

Although JSDMs are receiving increasing attention, there has been a lack of clarification on both the ecological processes they incorporate and on their specific commonalities and advantages with respect to SDMs. Since JSDMs infer a correlation matrix from the residual, it is tempting to think these residual correlations can inform about biotic interactions (Pollock *et al.* 2014) or even that JSDMs "account for biotic interactions in species distribution models" (Wilkinson *et al.* 2019). As highlighted by Poggiato *et al.* (2021), these tempting ideas should be avoided. JSDMs can provide additional information on species co-occurrences, but cannot separate the biotic and the abiotic effects, and their predictions on species distribution inevitably coincide with those of SDMs. However, JSDMs have the great advantage of leveraging on the residual correlation matrix to provide conditional predictions, which can be of great help in empirical studies, as we show in the case study below.

7.7. Case study

7.7.1. *Introduction of the dataset*

We present hereafter an application of latent factor models to a plant community dataset recently published by Martinez-Almoyna *et al.* (2020). The data are being collected within ORCHAMP, a long-term observatory of mountain ecosystems aiming to observe, understand and model biodiversity and ecosystem functioning over space and time. ORCHAMP is built around multiple elevational gradients that range from about 900 to 3000 m, and have been chosen to have a homogeneous exposure and slope along the gradient, a typical vegetation for the elevation levels (with woods dominating the lower parts and alpine meadows the higher parts), so that all the gradients as a whole are representative of the environmental and topographical variability of the French Alps. Between 2016 and 2018, at least five sampling plots were installed along 18 gradients, with an average of 200 m elevation difference, for a total of 99 plots (Figure 7.2). Here, we study the response of plant species to climate, the physicochemistry properties and the microbial activities of the soil. We applied latent factor models to a selection of 44 plant species, whose occurrences were recorded in at least 20 sites over the 99 sites, together with climatic variables, soil physicochemical properties and exoenzymatic activities. Latent factor models are particularly suitable to study the response of plant communities for the reasons described above. We aim to understand which species share the same response to the environment, and how eventual changes of climate and soil could affect these plants. Moreover, we are interested in the inference of the residuals correlation among species, the correlation matrix $\mathbf{\Sigma}$ that is given by $\mathbf{\Lambda}t(\mathbf{\Lambda})$. Because of the latent factor representation, we will be able not only to infer the residual associations among species, but also to represent species and sites on ordination axes, after filtering from the environment.

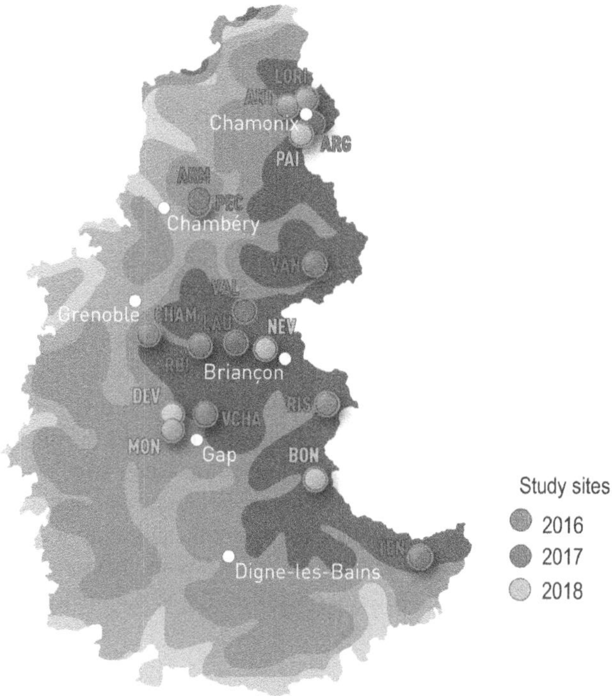

Figure 7.2. *Localization and names of the 18 gradients of ORCHAMP. For a color version of this figure, see www.iste.co.uk/peyrard/ecology.zip*

Using this dataset, Martinez-Almoyna *et al.* (2020) highlighted how Growing Degree Days (GDD, the annual sum of average daily degrees above zero), the total potential exoenzymatic activity (total EEA, the sum of all measured exoenzyme activities), soil pH and the ratio between soil carbon and nitrogen (soil C/N) determine the distributions of the 44 plant species. We therefore chose to include these four variables as the covariates of the latent factor model. In line with (Martinez-Almoyna *et al.* 2020), we considered the square of the GDD, due to the unimodal response of species to this variable.

7.7.2. *R package used*

To analyze the dataset, we used the R package Hmsc (Tikhonov *et al.* 2019, 2020). This package makes inference on the parameters of the models by sampling from the posterior distribution through Markov chain Monte Carlo (MCMC) sampling. Hmsc implements the latent factors methodology of Bhattacharya and Dunson (2011), where the number of latent factors is automatically chosen via shrinkage. Although we shall not describe all the features of this package, let us mention the interesting feature that it allows hierarchical modeling of the regression coefficients, and allows both functional traits and phylogeny to be included. This feat enables the user to study the dependence between functional traits and the environment, and to quantify the importance of phylogeny on species distribution. Moreover, it allows an explicit spatial and temporal dependence between sites to be included, improving the performance of the model. Here, however, we do not include any of these features to strictly describe the application of latent factor models[1].

7.7.3. *Implementation and convergence diagnosis*

We run two MCMC chains of 1,500 samples each, with 500 burn-in iterations and no thinning. These models are usually computationally demanding, and the computations for this model notably took around 3 h. Figure 7.3 shows that all the models clearly converged. The effective sample size (ESS) of the chains is very high for most parameters, and the potential scale reduction factor (psrf) was always close to one (the description of these measures can be found in Gelman *et al.* 2013). Thanks to the Bayesian framework, the full posterior distribution of the parameters was available and could then be used to compute a point estimate (posterior mean) and credible intervals (through posterior quantiles) for all parameters.

7.7.4. *Results and discussion*

We evaluated the predictive performance of the model both in in-sample prediction and in cross-validation (due to the high computational costs, we performed a twofold cross-validation only). We evaluated the model on these tasks by calculating, for each species, the true skill statistic (TSS), which has the advantage to account both for the model sensitivity (i.e. proportion of observed presences predicted as presences) and specificity (i.e. the proportion of observed absences predicted as absences; (Allouche *et al.* 2006)). TSS can vary from -1 to 1, where $+1$ indicates perfect fit and values of zero or less indicate a performance no better or worse than random (Allouche *et al.* 2006). Since the TSS requires a threshold to

1 The R code can be found at https://oliviergimenez.github.io/code_livre_variables_cachees/bystrova.html.

transform species' probability of presence into binary presence–absence data, we selected the threshold that maximizes the TSS values. We also evaluated the root mean square error (RMSE) of each species. In general, the model has good abilities to fit the data (mean in-sample TSS is equal to 0.63, Figure 7.4), but a scarcer ability to generalize on new data (in cross-validation the mean TSS drops by 0.5 and RMSE increases by 0.25). Model performances vary across species, with some species that were poorly modeled (three species had a cross-validation TSS score equal to 0) and others whose distribution was very explained (cross-validation TSS over 0.3).

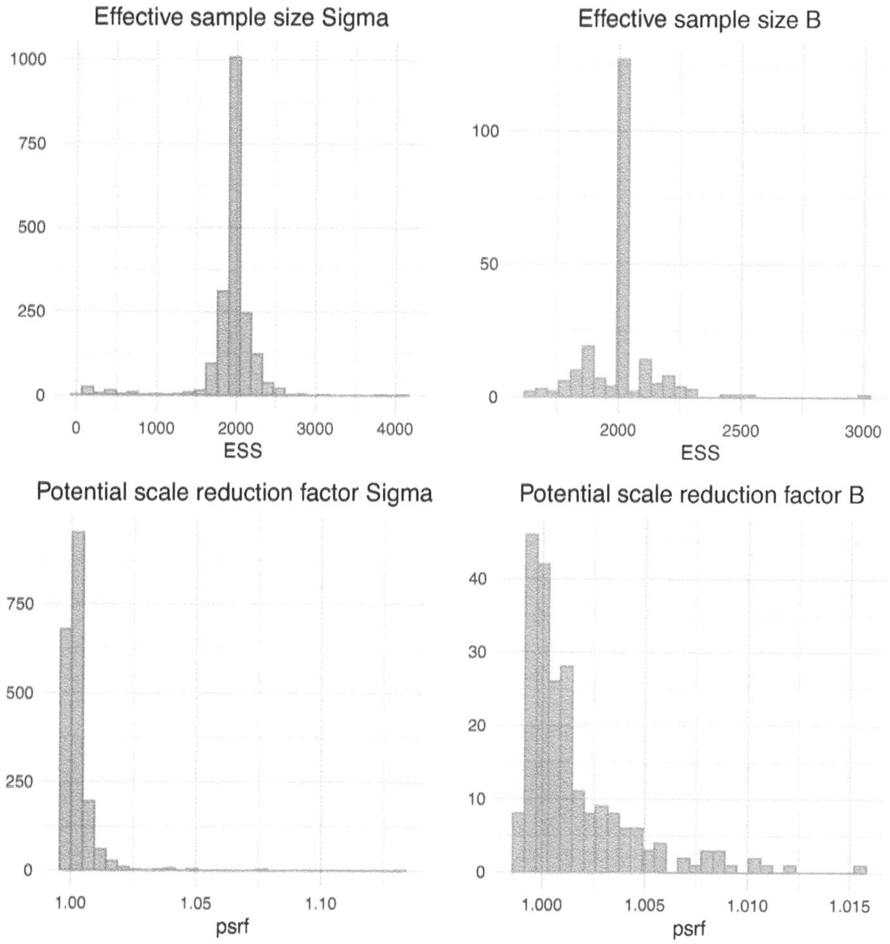

Figure 7.3. *Effective sample size (ESS, top panels) and potential scale reduction factor (psrf, bottom panels) for the correlation matrix Σ (Sigma, left panels) and the regression coefficient β (B, right panels). For a color version of this figure, see www.iste.co.uk/peyrard/ecology.zip*

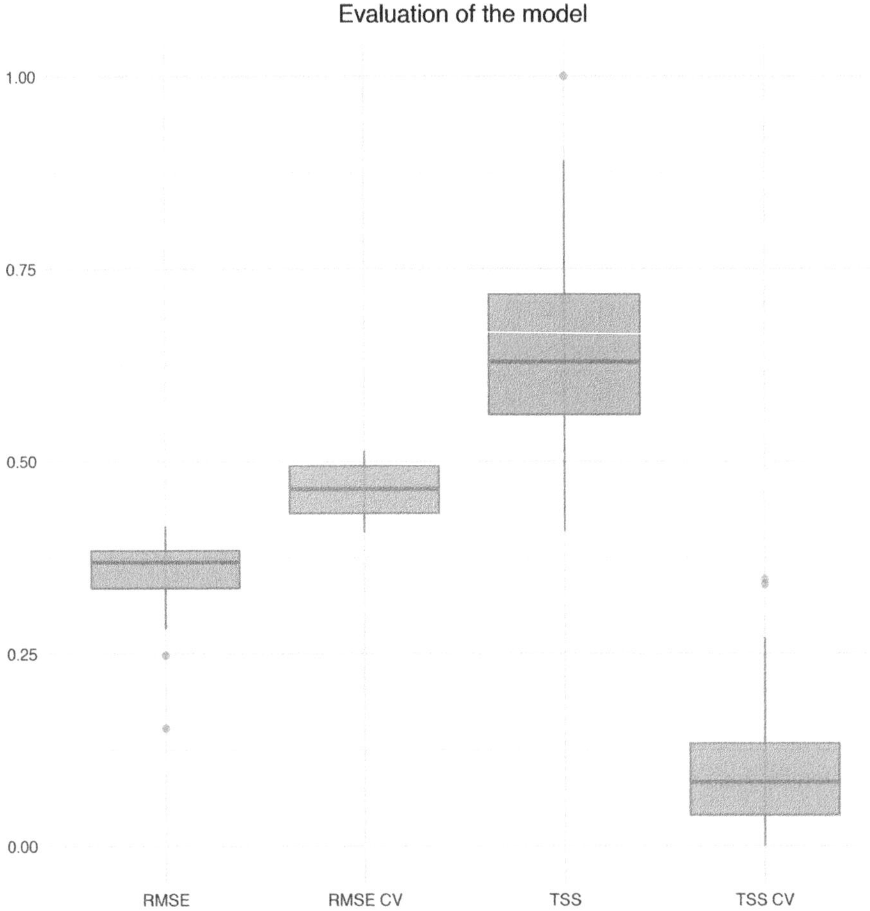

Figure 7.4. *Distribution of TSS and RMSE score across species for in-sample prediction (red) and twofold cross-validation (blue). For a color version of this figure, see www.iste.co.uk/peyrard/ecology.zip*

The regression coefficients tell us how species respond to the environment, and in this example, their heterogeneity shows how plant species have different responses to the environment (Figure 7.5). In general, climate (represented by GDD) has a significant effect on the distribution of a few number of species only. Instead, soil properties had a higher explanatory power. Many species notably show a trade-off along the gradients of soil characteristics: species that have a positive response to soil C/N, often have a negative one to soil pH and/or total EEA and vice versa (Figure 7.5). These results are consistent with Martinez-Almoyna *et al.* (2020), where the authors, who also considered functional traits but not residual correlation, showed how such a behavior reflects the functional trade-offs between conservatives and exploitative species. Exploitative plants are advantaged in nutrient-rich places with mild climate, while conservative species succeed in places where soil conditions are harsh because their adaptations allow them to survive in stressful situations. As a concluding remark, note that some species do not respond significantly to any of the environmental covariates, and these are the same species for which the TSS and RMSE scores are particularly poor. By analyzing the residual correlation matrix, we can understand species co-occurrence patterns that are not described by the environment, and provide insights about the phenomena that generate them. In the residual correlation matrix of this case study, the species are, interestingly, divided into two groups. Most plants tend to be positively correlated with species belonging to the same group, but negatively correlated with those of the other group (Figure 7.6). One group (that contains most of the species) is characterized by herbaceous plants that characterize alpine meadows (e.g. *Festuca violacea*, *Sesleria caerula*, *Carex curvula* and *Gentiana acaulis*), while the other one contains trees (e.g. *Picea Abies*), shade-preferring shrubs (e.g. *Vaccinium myrtillus* and *Vaccinium vitis-ideae*) and herbaceous species that are found in forests and humid habitats (e.g. *Melampyrum sylvaticum* and *Chaerophyllum villarsii*).

This residual correlation matrix highlights ecological phenomena that are well recognized. In fact, along elevational gradients, trees are limited by climatic conditions that prevent their survival above certain altitudes. As a result, herbaceous plants that need a high amount of light are excluded from the forests and are only found in open habitats, whereas other herbaceous plants (and shade-friendly shrubs) need the shade provided by the trees, and are therefore found in closed and/or humid habitats. The residual co-occurrence matrix not only endorses the biotic phenomena we described above, but also suggests to include habitat as an additional covariate, that might explain some of this residual co-occurrence patterns and improve model predictions.

Thanks to the latent factor representation, we can try to better understand where these correlations come from. A natural way of doing this is via *ordination*, as explained in section 7.5. We project the species in the space of the first two latent loadings (the first two columns of $\mathbf{T}$) and the sites in the first two latent factors thanks to a biplot (Figure 7.7). With such representation, we can think of latent factors as missing covariates, and represent sites depending on these missing covariates. Species loadings are therefore the response of species to such missing covariates. If two species are close on the biplot and far from the origin, they respond in the same way to these missing covariates, and are thus more correlated. For example, we see that *Picea abies*, *Melampyrum sylvaticum* and *Vaccinium myrtillus* tend to respond differently from the other species to these missing variables, and in fact, as said above, they are negatively correlated with them.

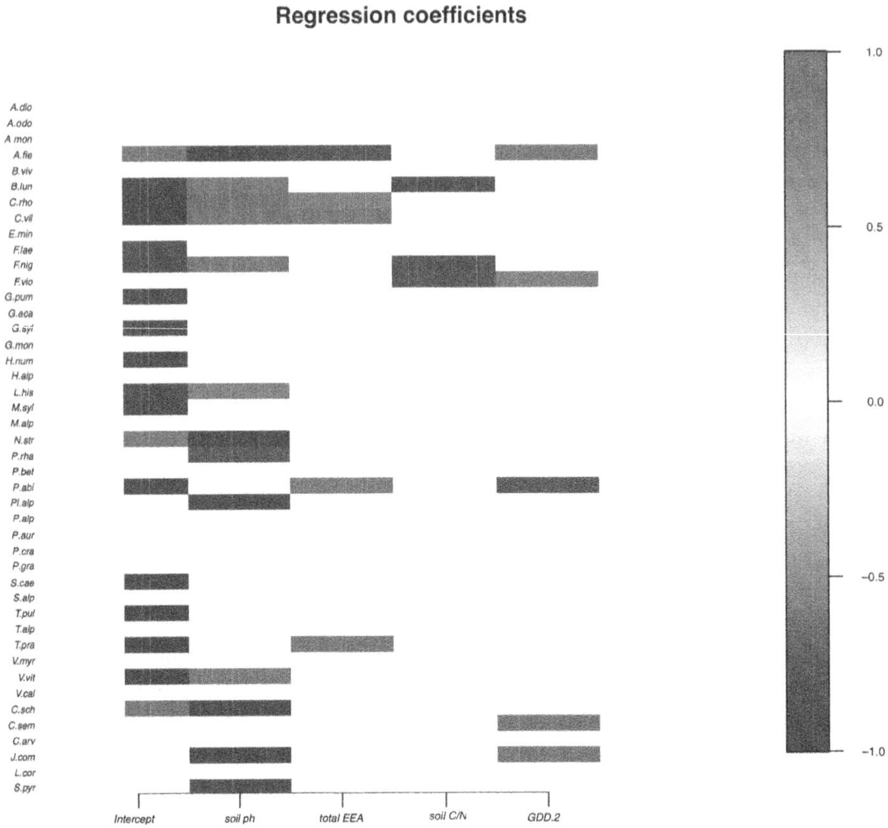

Figure 7.5. *Posterior support values for species regression coefficients. Red if the bounds of the 90% credible interval are both positive, white if the credible interval overlaps 0 and blue if both bounds are negative. For a color version of this figure, see www.iste.co.uk/peyrard/ecology.zip*

The type of habitat of the sites is one of the environmental predictors that we have not included in the study, and that could potentially impact species distribution, and interestingly some of the species highlighted in the previous biplot (Figure 7.7) tend to prefer close rather than open habitats compared to other species. We therefore marked each site as forest (indexed by 1) or grassland (0), re-run the model including this additional covariate and analyze its updated ordination plot (Figure 7.8). The above-mentioned species, which tend to behave as outliers when the habitat information was not include, are now closer to the rest of species. The species pool now tend to be more evenly distributed in the ordination space, even if some trends are still remarkable. Notably, we can still see a gradient, with sub-alpine species in the upper-right corner of the biplot and alpine species in the bottom-left corner. If this can this be still partially due to some unmeasured environmental variable, this might also be due to the influence of species on each others, with *Picea Abies* that provides the shade for other species.

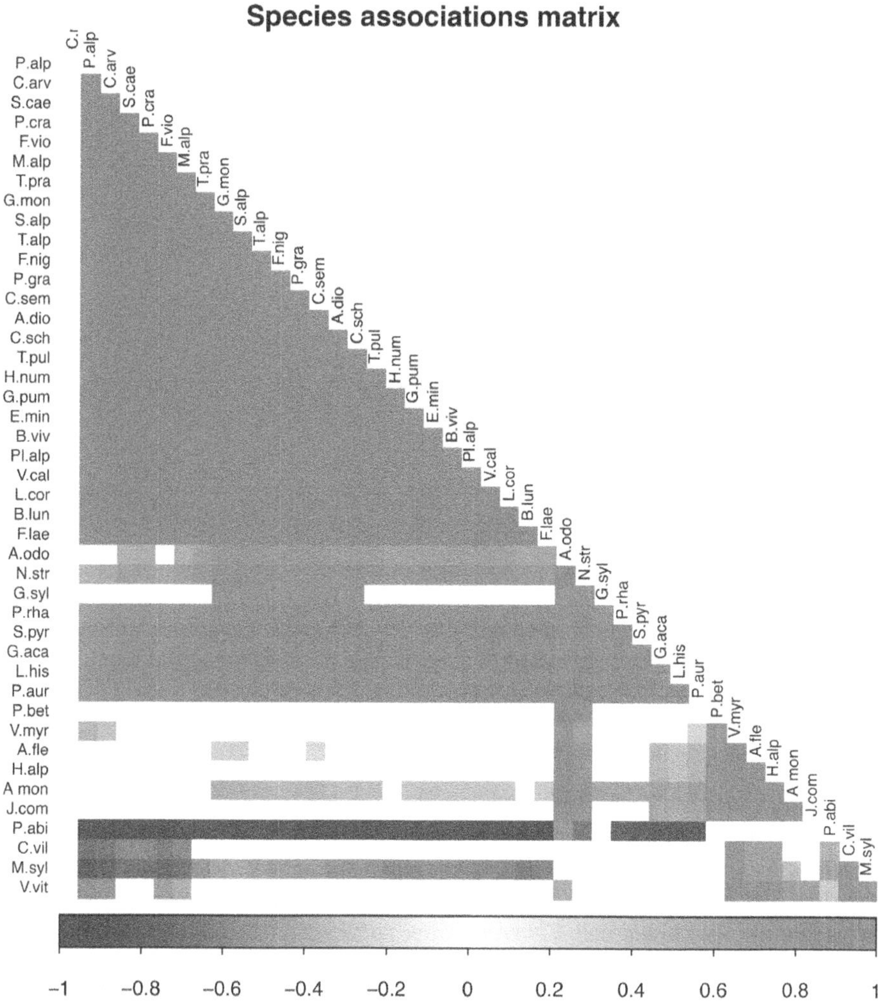

Figure 7.6. *The residual correlation matrix. Only significant values (i.e.* 95% *credible interval do not overlap zero) are shown. For a color version of this figure, see www.iste.co.uk/peyrard/ecology.zip*

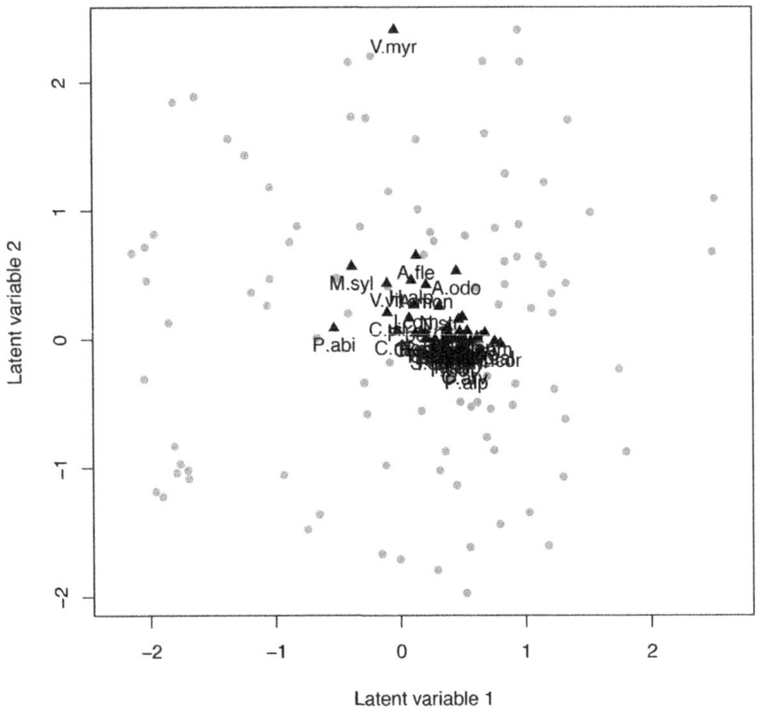

Figure 7.7. *Model-based ordination analysis. The two latent variables can be seen as missing covariates, and the position of species (black triangles) on the plot show the way species respond to those missing covariates. Species close in the latent variable species are positively correlated, and vice versa*

Finally, we want to build on the information that we assessed in the residual correlation matrix to improve the predictions of the model. We saw how *Picea abies* provides the shade and moisture that allows shrub species like *Vaccinium* (*Vaccinium myrtillus* and *Vaccinium vitis-ideae*) and shadow-friendly herbaceous species such as *Melampyrum sylvaticum* and *Chaerophyllum villarsii*) to thrive, while at the same time preventing the survival of herbaceous species that need lots of light, such as *Festuca violacea*. To improve our ability to predict one (or more) of the species described above, we predict the probability of occurrence of species conditionally on the presence (or absence) of *Festuca violacea*, a herbaceous plants that characterizes alpine grasslands. This is very similar to include *Festuca violacea* as predictor for the unobserved species. While including the other species as predictors is a doable option for communities with small number of species, it is not straightforward to do it if there are tens or hundreds of species. In contrast, conditional prediction can be made also for a great number of species, without the need to run the model again. When conditioning on *Festuca violacea*, the predictive power of the model improves, in particular concerning cross-validation predictions, where the mean TSS score gains 80% (from 0.1 to 0.19) with respect to the non-conditional predictions. This is particularly true for species that show a particular residual correlation (negative or positive) with *Festuca violacea*. Therefore, we focused on *Poa Alpina*, *Campanula scheuchzeri*, *Soldanella alpina*, *Viola calcarata* and *Euphrasia minima*, which, like *Festuca violacea*, characterize sub-alpine pastures and are often found together, and on the tree *Picea abies*, which as mentioned before takes the light that would allow the *Festuca violacea* to survive. For example, we consider an alpine meadow in the region of Devoluy (south of France), at an altitude of 2,100 m, where all the above-mentioned herbaceous species are present and *Picea abies* is absent. Figure 7.9a shows how cross-validation predictions conditioned by the presence of *Festuca violacea* increase the probability of the species that are actually present, and decrease the one of the tree, which is absent. This leads more generally to a marked improvement in the cross-validation AUC (Figure 7.9b).

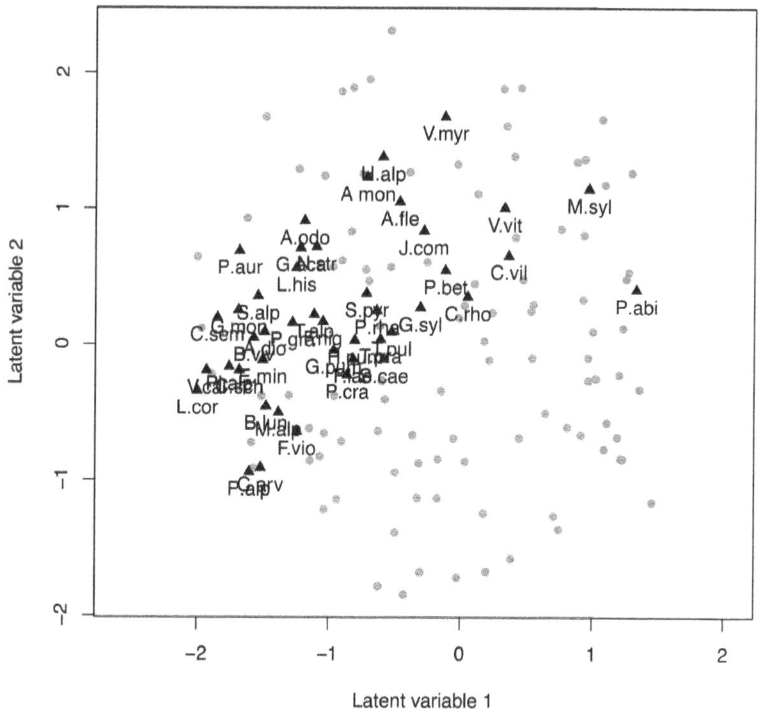

Figure 7.8. *Model-based ordination analysis, as above, but when we include the habitat as an additional covariate of the model*

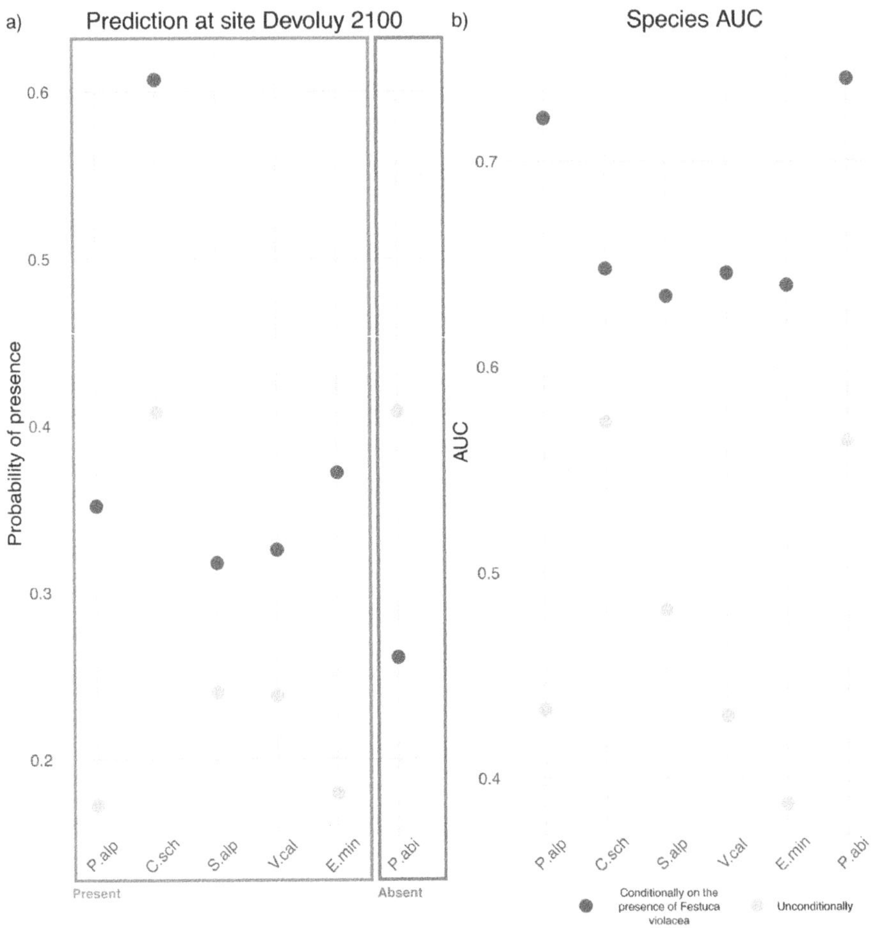

Figure 7.9. *Cross-validation predicted probability of (a) presence and (b) cross-validation AUC of Poa Alpina, Campanula scheuchzeri, Soldanella alpina, Viola calcarata, Euphrasia minima and Picea Abies conditionally on Festuca violacea (green) and unconditionally (yellow). At site Devoluy 2100 all the herbaceous species of above were present (green box) while Picea abies was absent (red box). For a color version of this figure, see www.iste.co.uk/peyrard/ecology.zip*

7.8. Conclusion

JSDMs have been recently proposed as an extension of SDMs that infers residual correlations between species, reflecting co-occurrence patterns not explained by the environmental predictors. These models should be interpreted with care (Poggiato *et al.* 2021), but they still provide important insights on community assemblage processes. In particular, the application of latent factors to JSDMs can provide further advantages. Indeed, latent factors reduce the dimension of the residual covariance matrix, and the related computational costs that were one of the strongest limitations of early JSDMs. Moreover, by measuring the main axes of residual co-variation between species, they also allow for a residual model based ordination of species and sites. This is particularly interesting when one aims to study species response to missing environmental variables, which is naturally measured by latent variables. Nevertheless, considering latent factors instead of a full residual covariance matrix can have some drawbacks. First, latent factor models increase their dimension with the number of sites. As a result, when dealing with many sites and few species, it is computationally more interesting to model a full residual covariance matrix. Moreover, it is not possible to sparsify the residual covariance matrix induced by latent factor models, a feature that has just been proposed as a solution to improve the interpretability of JSDMs (see Pichler and Hartig 2021) in the case of the full residual covariance matrix.

JSDMs have been implemented in many R packages, each with its particular features (see Wilkinson *et al.* 2019 for a review). In our case study, we chose to work with Hmsc because of its broad documentation and the large number of options it includes. Among them, it allows to take into account functional traits and phylogeny, and easily computes conditional predictions. Hmsc is a complete package, easy to start working with, but it is computationally heavy. In order to have faster results, we suggest working with the package proposed by Pichler and Hartig (2021), the features of which remain quite limited for now.

7.9. References

Allouche, O., Tsoar, A., Kadmon, R. (2006). Assessing the accuracy of species distribution models: Prevalence, kappa and the true skill statistic (TSS). *Journal of Applied Ecology*, 43(6), 1223–1232.

Barve, N., Barve, V., Jiménez-Valverde, A., Lira-Noriega, A., Maher, S.P., Peterson, A.T., Soberon, J., Villalobos, F. (2011). The crucial role of the accessible area in ecological niche modeling and species distribution modeling. *Ecological Modelling*, 222(11), 1810–1819 [Online]. Available at: http://www.sciencedirect.com/science/article/pii/S0304380011000780.

Bhattacharya, A. and Dunson, D.B. (2011). Sparse Bayesian infinite factor models. *Biometrika*, 98(2), 291–306.

Bystrova, D., Poggiato, G., Bektas, B., Arbel, J., Clark, J.S., Guglielmi, A., Thuiller, W. (2021). Clustering species with residual covariance matrix in joint species distribution models. *Frontiers in Ecology and Evolution*, 9, 128.

Calabrese, J.M., Certain, G., Kraan, C., Dormann, C.F. (2014). Stacking species distribution models and adjusting bias by linking them to macroecological models. *Global Ecology and Biogeography*, 23(1), 99–112 [Online]. Available at: https://onlinelibrary.wiley.com/doi/abs/10.1111/geb.12102.

Chib, S. and Greenberg, E. (1998). Analysis of multivariate probit models. *Biometrika*, 85, 347–361.

Clark, J.S. and Gelfand, A.E. (2006). *Hierarchical Modelling for the Environmental Sciences: Statistical Methods and Applications*. Oxford University Press, Oxford.

Clark, J.S., Gelfand, A.E., Woodall, C.W., Zhu, K. (2014). More than the sum of the parts: Forest climate response from joint species distribution models. *Ecological Applications*, 24(5), 990–999.

Clark, J.S., Nemergut, D., Seyednasrollah, B., Turner, P., Zhang, S. (2017). Generalized joint attribute modeling for biodiversity analysis: Median-zero, multivariate, multifarious data. *Ecological Monographs*, 87, 34–56.

Ellison, A.M. (2004). Bayesian inference in ecology. *Ecology Letters*, 7(6), 509–520.

Gelman, A., Carlin, J.B., Stern, H.S., Dunson, D.B., Vehtari, A., Rubin, D.B. (2013). *Bayesian Data Analysis*, 3rd edition. Chapman and Hall/CRC, New York.

Guisan, A. and Rahbek, C. (2011). Sesam: A new framework integrating macroecological and species distribution models for predicting spatio-temporal patterns of species assemblages. *Journal of Biogeography*, 38(8), 1433–1444.

Guisan, A. and Thuiller, W. (2005). Predicting species distribution: Offering more than simple habitat models. *Ecology Letters*, 8(9), 993–1009 [Online]. Available at: https://onlinelibrary.wiley.com/doi/abs/10.1111/j.1461-0248.2005.00792.x.

Guisan, A. and Zimmermann, N.E. (2000). Predictive habitat distribution models in ecology. *Ecological Modelling*, 135(2), 147–186 [Online]. Available at: http://www.sciencedirect.com/science/article/pii/S0304380000003549.

Guisan, A., Thuiller, W., Zimmermann, N.E. (2017). *Habitat Suitability and Distribution Models: With Applications in R*. Cambridge University Press, Cambridge.

Hutchinson, G.E. (1957). Concluding remarks. *Cold Spring Harbor Symposia on Quantitative Biology*, 22, 415–427.

Lortie, C.J., Brooker, R.W., Choler, P., Kikvidze, Z., Michalet, R., Pugnaire, F.I., Callaway, R.M. (2004). Rethinking plant community theory. *Oikos*, 107(2), 433–438 [Online]. Available at: https://onlinelibrary.wiley.com/doi/abs/10.1111/j.0030-1299.2004.13250.x.

Martinez-Almoyna, C., Piton, G., Abdulhak, S., Boulangeat, L., Choler, P., Delahaye, T., Dentant, C., Foulquier, A., Poulenard, J., Noble, V., Renaud, J., Rome, M., Saillard, A., Consortium, T.O., Thuiller, W., Münkemüller, T. (2020). Climate, soil resources and microbial activity shape the distributions of mountain plants based on their functional traits. *Ecography*, 43(10), 1550–1559 [Online]. Available at: https://onlinelibrary.wiley.com/doi/abs/10.1111/ecog.05269.

McCullagh, P. and Nelder, J. (1989). *Generalized Linear Models*, 2nd edition. Chapman & Hall, Boca Raton [Online]. Available at: http://books.google.com/books?id=h9kFH2_FfBkC.

Merow, C., Smith, M.J., Edwards Jr, T.C., Guisan, A., McMahon, S.M., Normand, S., Thuiller, W., Wüest, R.O., Zimmermann, N.E., Elith, J. (2014). What do we gain from simplicity versus complexity in species distribution models? *Ecography*, 37(12), 1267–1281 [Online]. Available at: https://onlinelibrary.wiley.com/doi/abs/10.1111/ecog.00845.

Ovaskainen, O. and Abrego, N. (2020). *Joint Species Distribution Modelling: With Applications in R*. Cambridge University Press, Cambridge.

Ovaskainen, O. and Soininen, J. (2011). Making more out of sparse data: Hierarchical modeling of species communities. *Ecology*, 92(2), 289–295 [Online]. Available at: https://esajournals.onlinelibrary.wiley.com/doi/abs/10.1890/10-1251.1.

Ovaskainen, O., Tikhonov, G., Norberg, A., Guillaume Blanchet, F., Duan, L., Dunson, D., Roslin, T., Abrego, N. (2017). How to make more out of community data? A conceptual framework and its implementation as models and software. *Ecology Letters*, 20(5), 561–576 [Online]. Available at: https://onlinelibrary.wiley.com/doi/abs/10.1111/ele.12757.

Pichler, M. and Hartig, F. (2021). A new joint species distribution model for faster and more accurate inference of species associations from big community data. *Methods in Ecology and Evolution*, 12(11), 2159–2173.

Poggiato, G., Münkemüller, T., Bystrova, D., Arbel, J., Clark, J., Thuiller, W. (2021). On the interpretations of joint modelling in community ecology. *Trends in Ecology and Evolution*, 36(5), 391–401.

Pollock, L.J., Tingley, R., Morris, W.K., Golding, N., O'Hara, R.B., Parris, K.M., Vesk, P.A., McCarthy, M.A. (2014). Understanding co-occurrence by modelling species simultaneously with a joint species distribution model. *Methods in Ecology and Evolution*, 5(5), 397–406.

Pulliam, H. (2000). On the relationship between niche and distribution. *Ecology Letters*, 3(4), 349–361 [Online]. Available at: https://onlinelibrary.wiley.com/doi/abs/10.1046/j.1461-0248.2000.00143.x.

Soberon, J. (2007). Grinnellian and Eltonian niches and geographic distributions of species. *Ecology Letters*, 10(12), 1115–1123 [Online]. Available at: https://onlinelibrary.wiley.com/doi/abs/10.1111/j.1461-0248.2007.01107.x.

Soberon, J. and Peterson, A. (2005). Interpretation of models of fundamental ecological niches and species distributional areas. *Biodiversity informatics*, 2, 1–10.

Spiegelhalter, D.J., Best, N.G., Carlin, B.P., Van Der Linde, A. (2002). Bayesian measures of model complexity and fit. *Journal of the Royal Statistical Society: Series B (Statistical Methodology)*, 64(4), 583–639.

Taberlet, P., Coissac, E., Hajibabaei, M., Rieseberg, L.H. (2012). Environmental DNA. *Molecular Ecology*, 21(8), 1789–1793.

Thuiller, W., Münkemüller, T., Lavergne, S., Mouillot, D., Mouquet, N., Schiffers, K., Gravel, D. (2013). A road map for integrating eco-evolutionary processes into biodiversity models. *Ecology Letters*, 16(s1), 94–105 [Online]. Available at: https://onlinelibrary.wiley.com/doi/abs/10.1111/ele.12104.

Tikhonov, G., Ovaskainen, O., Oksanen, J., de Jonge, M., Opedal, O., Dallas, T. (2019). *HMSC: Hierarchical Model of Species Communities.* R package version 3.0-4 [Online]. Available at: https://CRAN.R-project.org/package=Hmsc.

Tikhonov, G., Opedal, Ø.H., Abrego, N., Lehikoinen, A., de Jonge, M.M.J., Oksanen, J., Ovaskainen, O. (2020). Joint species distribution modelling with the R-package HMSC. *Methods in Ecology and Evolution*, 11(3), 442–447.

Warton, D., Foster, S.D., De'ath, G., Stoklosa, J., Dunstan, P.K. (2015). Model-based thinking for community ecology. *Plant Ecology*, 216, 669–682.

Watanabe, S. and Opper, M. (2010). Asymptotic equivalence of Bayes cross validation and widely applicable information criterion in singular learning theory. *Journal of Machine Learning Research*, 11(116), 3571–3594.

Wilkinson, D.P., Golding, N., Guillera-Arroita, G., Tingley, R., McCarthy, M.A. (2019). A comparison of joint species distribution models for presence–absence data. *Methods in Ecology and Evolution*, 10(2), 198–211.

Yates, K., Bouchet, P., Caley, M., Mengersen, K. (2018). Outstanding challenges in the transferability of ecological models. *Trends in Ecology & Evolution*, 33(10), 790–802 [Online]. Available at: http://usir.salford.ac.uk/id/eprint/47980/.

8

The Poisson Log-Normal Model: A Generic Framework for Analyzing Joint Abundance Distributions

Julien CHIQUET[1], Marie-Josée CROS[2], Mahendra MARIADASSOU[3], Nathalie PEYRARD[2] and Stéphane ROBIN[1]
[1] Paris-Saclay University, AgroParisTech, INRAE, UMR MIA-Paris, France
[2] University of Toulouse, INRAE, UR MIAT, Castanet-Tolosan, France
[3] Paris-Saclay University, INRAE, MaIAGE, Jouy-en-Josas, France

8.1. Introduction

The way in which an ecosystem works is essentially dependent on the interactions between the species making up the system and their environment (abiotic interactions) and on interactions between these species (biotic interactions). By characterizing the diversity of these communities, we gain the ability to monitor their evolution over time, and/or to understand how observed patterns in a community may vary depending on the environment. From a conservation (or monitoring) perspective, this approach also offers a means of evaluating the effects of protection measures and defining targeted actions (Tylianakis *et al.* 2010; Xiao *et al.* 2018).

The diversity of ecosystems can be studied at varying levels, from the microscopic, for example in the case of gut microbiota (Layeghifard *et al.* 2017), to the macroscopic, for example communities of trees in forests (Clark *et al.* 2014). In both cases, the initial data take the form of presence–absence tables, or a count of individuals of different species observed in different samples, characterized by different environmental properties. In microbiology, metabarcoding data are used to provide a count of "*reads*" (sequences), while in forest or marine ecology, individuals may be counted directly.

Simply looking at the number or variety of species in a community, or even its Shannon index, is not enough to give us a fine description of these ecosystems. These summary values do not provide information concerning the way in which species collectively respond to their environment, or about associations (positive or negative) between species.

Statistical models have been designed to respond to these limits, focusing on associations between species and their joint reaction to the environment. In this context, the presence–absence (Harris 2015; Ovaskainen *et al.* 2017) or abundance (Popovic *et al.* 2018) of all the species in question are modeled jointly. The resulting models are known as joint species distribution models (JSDM) and are notably different from species abundance models (SDM: (Elith and Leathwick 2009)), which consider the influence of the environment on the abundance of a single species. The application of JSDM to presence–absence data was discussed in Chapter 7; in this chapter, we shall consider the case of count data.

From a statistical perspective, the joint modeling of abundance data presents a number of problems due to the very nature of the data, notably the fact that (i) these data consist of counts and (ii) the observed distribution is often overdispersed compared to a Poisson distribution, which serves as a reference for count data. Concerning point (i), note that in the case of continuous data, the normal multivariate distribution is used, but there is no natural multivariate distribution for count data. Multivariate models for count data do exist, but these often come with strict constraints, notably in terms of the sign, with respect to dependency relationships between species (Inouye *et al.* 2017).

Given these issues, JSDM modelers have naturally turned toward latent variable models (Warton *et al.* 2015), which offer greater flexibility in terms of modeling dependency. This approach notably enables count data to be used in the same way as continuous data within the latent variables. Several of these models use a Gaussian distribution for the latent layer (Ovaskainen *et al.* 2017; Popovic *et al.* 2018): dependencies between species are described by the covariance matrix of the latent vector associated with each sample. The Poisson log-normal model (PLN: Aitchison and Ho (1989)) discussed in this chapter falls within this framework. One advantage of latent model variables lies in the fact that they induce overdispersion by construction, simply by including an additional random element in the distribution of observations. Finally, these models make it relatively easy to take account of environmental effects, that is, abiotic interactions, on the abundance of different species, by means of regression.

In this chapter, we shall illustrate the possibilities offered by the PLN model using marine species count data collected by the PISCO research program (PISCO Research Consortium 2019b). Coastal marine ecosystems are currently subject to severe disturbances (overfishing, habitat destruction, pollution, etc.) along with the effects of climate change, which include both ecological consequences (extreme temperatures, acidification, invasion of new species) and socioeconomic consequences (Pan *et al.* 2013). Developing a clearer understanding of these ecosystems is key to protecting them and to managing the effects of human activities. The PISCO research program was launched in 1999, studying marine communities along the US west coast (PISCO Research Consortium 2019b). The aim of the program is to develop a better understanding of the causes and consequences of changes to the ecosystem. Activities include long-term species sampling, aimed at improving knowledge of species distribution and interactions.

One of the ecosystems featured in the study is the kelp forest (PISCO Research Consortium 2019a) in the Channel Islands archipelago off the coast of Santa Barbara (southern California). The creation of a national park around these islands means that they act as a refuge for marine life (Caselle 2013). For the purposes of this study, we have chosen to focus on one island, Anacapa (made up of three islets, or "sites" in this context): the surrounding area is relatively well protected, and the marine environment here has been monitored since 1999. We shall use the MariNet dataset, extracted from the PISCO data, to show how the PLN model may be used in response to three types of questions: evaluating the influence of covariates, such as site or year, on species abundance; identifying species which react to these covariates in the same way; and identifying direct interactions between species.

8.2. The Poisson log-normal model

8.2.1. *The model*

We shall begin by presenting the general Poisson log-normal model, including the way in which it takes account of both abiotic effects and biotic interactions, alongside sampling effort.

The multivariate Poisson log-normal (PLN) model, introduced by Aitchison and Ho (1989), is a latent variable model with the capacity to describe an overdispersed S-dimensional count vector. Consider a sample of size N created by independent drawings from such a vector. In the case of the MariNet data, S represents the number of species, and a sample is defined by four elements, including the site and the year (see section 8.3). The PLN model links each observation $\mathbf{Y}_n = (Y_{n1}, \dots, Y_{nS}) \in \mathbb{N}^S$ ($1 \leq n \leq N$) in the count vector to a latent Gaussian vector $\mathbf{Z}_n \in \mathbb{R}^S$, in such a way that the coordinates of $\mathbf{Y}_n$ are drawn independently, conditional on $\mathbf{Z}_n$, according to a Poisson distribution:

$$\begin{array}{ll} \text{latent space} & \mathbf{Z}_n \sim \mathcal{N}(\boldsymbol{\mu}, \boldsymbol{\Sigma}), \\ \text{observation } Y_{ns} \mid Z_{ns} & \text{indep. } \mathbf{Y}_n \mid \mathbf{Z}_n \sim \mathcal{P}\left(\exp\{\mathbf{Z}_n\}\right). \end{array} \qquad [8.1]$$

The vector of means $\boldsymbol{\mu} \in \mathbb{R}^S$ corresponds to the main effects, while the variance–covariance matrix $\boldsymbol{\Sigma}$ describes the dependency structure between the S coordinates of vector $\mathbf{Z}_n$. In the case of abundance data, this is the dependency structure between species within the collected samples. The dependency structure of the PLN model is illustrated in Figure 8.1. Figure 8.2 shows a geometric view of the PLN model for a case involving two species.

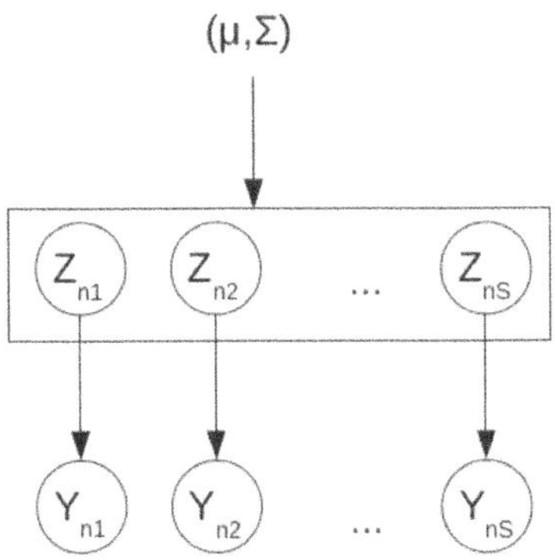

Figure 8.1. *Illustration of dependency in the PLN model. Random variables are circled, while parameters are not circled*

The PLN distribution is naturally overdispersed compared to a Poisson distribution, as we might expect in this context of application. If $\boldsymbol{\Sigma} = [\sigma_{sr}]_{1 \leq j,k \leq S}$, then $\mathbb{E}(Y_{ns}) = e^{\mu_s + \sigma_{ss}/2}$ and $\mathbb{V}(Y_{ns}) = \mathbb{E}(Y_{ns}) + (e^{\sigma_{ss}} - 1)\mathbb{E}(Y_{ns})^2 \geq \mathbb{E}(Y_{ns})$. Furthermore, the covariance between the observed counts of two species may take arbitrary signs: $\mathbb{C}\text{ov}(Y_{ns}, Y_{nr}) = (e^{\sigma_{sr}} - 1)\mathbb{E}(Y_{ns})\mathbb{E}(Y_{nr})$, and thus $\mathbb{C}\text{ov}(Y_{ns}, Y_{nr})$ has the same sign as $\mathbb{C}\text{ov}(Z_{ns}, Z_{nr}) = \sigma_{sr}$.

8.2.1.1. *Covariates and offsets*

Model [8.1] can be generalized naturally to give a formulation close to that of the general linear model (i.e. with multiple responses), in which the main effects correspond to a linear combination of D fixed covariates[1] noted $\mathbf{x}_n$. In the context of

1 In what follows, the covariate vector $\mathbf{x}_n$ also includes the intercept.

this study, it is also natural to include an offset matrix in the model, i.e. a fixed shift in the regression (known to the modeler) that depends on the sample, and potentially on the species. This notably makes it possible to model the notion of sampling effort, as we shall see later. Let $\mathbf{o}_n \in \mathbb{R}^S$ be the offset vector of sample n. Thus, model [8.1] can be generalized as

$$\mathbf{Y}_n \mid \mathbf{Z}_n \sim \mathcal{P}\left(\exp\{\mathbf{Z}_n\}\right), \qquad \mathbf{Z}_n \sim \mathcal{N}(\mathbf{o}_n + t(\mathbf{x}_n)\mathbf{B}, \Sigma), \qquad [8.2]$$

where $\mathbf{B}$ is the $D \times S$ matrix of regression coefficients.

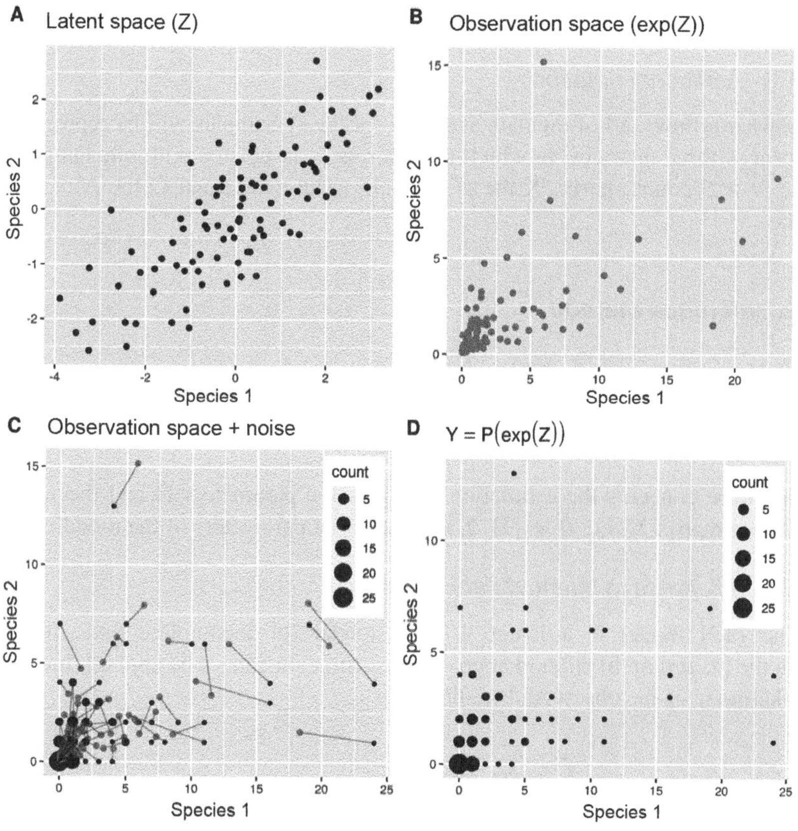

Figure 8.2. *PLN: geometric view of the model for two species. A) Positions in the latent space: mean log-abundances. B) Mean counts. C) Mean counts (red) and observed counts (black). D) Observed counts. For a color version of this figure, see www.iste.co.uk/peyrard/ecology.zip*

8.2.1.2. *Modeling the variance-covariance matrix*

The parametrization used to describe the variance–covariance matrix can be adjusted where necessary, notably in order to reduce the total number of parameters in the model. According to the most general hypothesis, matrix $\mathbf{\Sigma}$ has $S(S+1)/2$ parameters (S variance parameters and $S(S-1)/2$ covariate terms). However, the model designer may choose to describe the variances of species alone, using a diagonal matrix $\mathbf{\Sigma}$ with only S parameters. In this case, the model is equivalent to S independent SDMs. In extreme situations[2], a single variance parameter may be used for the whole matrix, such that $\mathbf{\Sigma} = \sigma \mathbf{I}_p$. Other types of model for matrix $\mathbf{\Sigma}$, suitable for dimension reduction and network inference, will be presented in sections 8.2.3 and 8.2.4.

8.2.1.3. *Additional notation*

In what follows, all of the data available for the N samples will be represented in the form of three matrices, in which the nth row is associated with nth sample, with $\mathbf{Y}$ the $N \times S$ count matrix, $\mathbf{X}$ the $N \times D$ covariate matrix and $\mathbf{O}$ the $N \times S$ offset matrix.

8.2.2. *Inference method*

In this section, we shall provide a brief description of ways of estimating the parameters of the PLN model, and of a number of difficulties typically encountered in the case of latent variable models.

Inference concerns the estimation of regression parameters $\mathbf{B}$ and the variance–covariance matrix $\mathbf{\Sigma}$. Let $\boldsymbol{\theta} = \{\mathbf{B}, \mathbf{\Sigma}\}$ be the set of parameters of the model.

8.2.2.1. *Inference in latent variable models*

The PLN model is a latent variable model (or incomplete data model) in which the maximum likelihood approach to estimation cannot be applied. Indeed, the log-likelihood of the observed data, that is,

$$\log p_{\boldsymbol{\theta}}(\mathbf{Y}) = \log \int_{\mathcal{Z}} p_{\boldsymbol{\theta}}(\mathbf{Y}, \mathbf{Z}) \mathrm{d}\mathbf{Z},$$

is impossible to estimate, due to the integration over the set $\mathcal{Z} = \mathbb{R}^S$ describing the space of possible values of the latent variable.

2 For example, when a series of PLN models are created using components of a mixture.

One common solution to this issue is to use the expectation–maximization algorithm (EM, Dempster *et al.* 1977) to maximize log-likelihood locally based on the conditional expectation of the log-likelihood of the complete data, that is,

$$\mathbb{E}_{\boldsymbol{\theta}}\left[\log p_{\boldsymbol{\theta}}(\mathbf{Y}, \mathbf{Z}) \mid \mathbf{Y}\right]. \qquad [8.3]$$

This approach can be used for estimation in a large number of latent variable models. Unfortunately, it cannot be used directly in a PLN model as the number of species S increases: in order to evaluate [8.3], we need to be able to integrate according to the distribution of each latent vector $\mathbf{Z}_n$, conditional on the count vector $\mathbf{Y}_n$. As this distribution does not have a closed form in the context of the PLN model, other numerical integration schemes or Monte Carlo approaches must be used; however, these are hard to apply for data concerning more than a few dozen species.

8.2.2.2. *Variational approximation*

An alternative approach is that of variational approximation, which consists of finding an approximate distribution for $p_{\boldsymbol{\theta}}(\mathbf{Z}_n \mid \mathbf{Y}_n)$ which simplifies the integration. The variational approach, as presented by Wainwright and Jordan (2008), consists of minimizing the Kullback–Leibler divergence KL between the actual conditional distribution and the approximate distribution, selected from a predefined class in order to simplify the calculation of [8.3]. In the case of PLN models, we propose to approximate $p_{\boldsymbol{\theta}}(\mathbf{Z}_n \mid \mathbf{Y}_n)$ by a multivariate Gaussian distribution denoted as q_n, with mean vector $\mathbf{m}_n$ and a diagonal variance–covariance matrix $\mathbf{S}_n = \text{diag}(\mathbf{s}_n^2)$. The set of variational parameters are collected in the vector $\psi = (\mathbf{M}, \mathbf{S})$, where $\mathbf{M} = t([t(\mathbf{m}_1) \ldots t(\mathbf{m}_n)])$, $\mathbf{S} = t([t(\mathbf{s}_1^2) \ldots t(\mathbf{s}_n^2)])$.

Using the Kullback–Leibler divergence to measure the quality of the approximation results in an approximate version of the EM algorithm (or variational EM), which maximizes a lower bound of the log-likelihood of the observations, defined by

$$\begin{aligned} J(\mathbf{Y}; \psi, \theta) &\triangleq \log p_{\boldsymbol{\theta}}(\mathbf{Y}) - KL\left[q_\psi(\mathbf{Z}) || p_{\boldsymbol{\theta}}(\mathbf{Z} \mid \mathbf{Y})\right] \\ &= \mathbb{E}_{q_\psi}\left[\log p_{\boldsymbol{\theta}}(\mathbf{Y}, \mathbf{Z})\right] - \mathbb{E}_{q_\psi}\left[\log q_\psi(\mathbf{Z})\right], \end{aligned} \qquad [8.4]$$

where $\mathbb{E}_{q_\psi}$ is the conditional expectation, under the variational distribution q_ψ. This approach may be generalized to all of the PLN variants presented in this chapter. The lower bound of the likelihood is optimized in ψ and $\boldsymbol{\theta}$ using a gradient ascent type approach: specifically, in this case, the CCSA algorithm designed by Svanberg (2002) and implemented in the C++ `nlopt` repository (Johnson 2011).

8.2.3. *Dimension reduction*

The first variant of the general PLN model presented here is particularly suitable for cases involving a large number of species, and is notably useful for visualizing large datasets.

8.2.3.1. *Presentation of the model*

In the general PLN model, the latent variable is assumed to belong to a latent space of the same dimension S as the observation space. This property results from the lack of constraints on the covariance matrix $\mathbf{\Sigma}$ of the law of latent vectors $\mathbf{Z}_n$. This hypothesis may be costly in cases involving a large number of species, and in such cases, one option is to consider that the Z_{ns} belong to a space of intrinsic dimension $K \ll S$. In the context of the PLN model, this can be done by defining the strict analog of the probabilistic principal component analysis (PCA) proposed by Tipping and Bishop (1999) in the Gaussian context. This variant is known as the PLNPCA model (Chiquet *et al.* 2018).

A first formulation of the PLNPCA model involves taking the matrix $\mathbf{\Sigma}$ to be of rank $K \ll S$. This hypothesis presumes the existence of a matrix $\mathbf{T}$ of dimension $S \times K$ such that

$$\mathbf{\Sigma} = \mathbf{T}t(\mathbf{T}). \qquad [8.5]$$

A more intuitive, alternative version of this formulation relies on the definition of a vector of latent factors $\mathbf{W}_n$ associated with each sample:

$$\begin{aligned} \text{latent vector space } \mathbf{W}_n &\sim \mathcal{N}(\mathbf{0}_K, \boldsymbol{I}_{K\times K}), \\ \text{latent space } \mathbf{Z}_n &= \boldsymbol{\mu} + \mathbf{T}\mathbf{W}_n \\ \Rightarrow \quad \mathbf{Z}_n &\sim \mathcal{N}(\boldsymbol{\mu}, \mathbf{T}t(\mathbf{T})). \end{aligned} \qquad [8.6]$$

Note that the latent vector $\mathbf{Z}_n$ is entirely determined by the vector $\mathbf{W}_n$: the PLNPCA does not, strictly speaking, induce an additional latent layer not present in the PLN model, defined in equation [8.1]. Just like the PLN model, the PLNPCA is flexible enough to include covariates, and this is done by modifying the distribution of latent variables defined in equation [8.2].

The set of parameters for the model is then $\boldsymbol{\theta} = \{\mathbf{B}, \mathbf{T}\}$, but note that the model remains unchanged if we replace $\mathbf{T}$ by $\mathbf{TO}$, where $\mathbf{O}$ is a rotation matrix of $\mathbb{R}^K$. In other terms, $\mathbf{T}$ is only identifiable to a rotation matrix, via the matrix $\mathbf{T}t(\mathbf{T})$, in the same way as in probabilistic PCA[3].

3 In this latter case, the orthogonality of axes and the sorting of axes in order of descending eigenvalue offers a solution to the non-identifiability problem.

8.2.3.2. *Model inference*

The process used to infer the PLNPCA model is the same as that used for the PLN model, but with two major differences. First, in the case of PLNPCA, the variational approximation is applied to $p_{\boldsymbol{\theta}}(\mathbf{W}_n \mid \mathbf{Y}_n)$. Vectors $\mathbf{m}_n$ and $\mathbf{s}_n$ are thus of size K, not S. Second, the number K of latent factors is generally unknown, so a model selection criterion must be used.

In practice, the inference method consists of the following steps: (i) selection of a maximum number of latent factors (noted $K_{\max}$), (ii) estimation of a PLNPCA model of size K for all values of K between 1 and $K_{\max}$ and (iii) selection of a value $\hat{K}$ which maximizes the following penalized likelihood criterion:

$$BIC(K) = J_K(\mathbf{Y}; \psi, \theta) - \frac{S(D+K)}{2},$$

where J_K is the variational likelihood [8.4] calculated for the PLNPCA model with K factors. This criterion can be modified to correspond to the BIC criterion (Schwarz 1978) by replacing the likelihood with its variational approximation.

8.2.3.3. *Using the results of the estimation process*

Once the parameters of the model have been estimated, we have an estimator $\widehat{\boldsymbol{\theta}}$ of $\boldsymbol{\theta}$ and an estimator $\widetilde{\psi}$ of ψ. These are used (i) to calculate the deviance explained by the model, (ii) to estimate the position of the samples in the latent space and (iii) to investigate the residual structure, that is, the part not explained by the covariates.

The deviance is calculated using the following formula:

$$D_{\hat{K}} = \frac{J_{\hat{K}}(\mathbf{Y}, \widetilde{\psi}, \widehat{\boldsymbol{\theta}}) - J_{\min}}{J_{\max} - J_{\min}},$$

where $J_{\max}$ is the likelihood of the saturated model, obtained by fixing $\mathbf{Z}_n = \log(\mathbf{Y}_{ns})$ in [8.4], and $J_{\min}$ is the likelihood of the model with 0 factors, obtained by fixing $\mathbf{Z}_n = \mathbf{o}_n + t(\mathbf{x}_n)\widehat{\mathbf{B}}$ in the same equation. The position of the samples in the latent space is given by $\widetilde{\mathbf{Z}_n} = \mathbf{o}_n + t(\mathbf{x}_n)\widehat{\mathbf{B}} + \widehat{\mathbf{T}}\widetilde{\mathbf{m}}_n$. For ease of visualization, we have chosen to focus on the term $\widehat{\mathbf{T}}\widetilde{\mathbf{m}}_n$ alone. This term corresponds to the *residual* structure, that is, what remains after correction for differences in sampling effort ($\mathbf{o}_n$) and for the effects of covariates ($t(\mathbf{x}_n)\widehat{\mathbf{B}}$). An illustration will be given in section 8.3.3.

8.2.4. *Inferring networks of interaction*

A second variant of the PLN model aims to reconstruct the network of ecological interactions, notably by distinguishing between statistical associations (correlations) and direct interactions between species pairs.

8.2.4.1. *PLN as a Gaussian graphical model*

The PLN model may be used to gain a clearer understanding of interactions between the species that make up an ecosystem. The covariance matrix $\mathbf{\Sigma} = [\sigma_{sr}]_{1\leq s,r\leq S}$ gives a first indication concerning these relations via the correlations $\rho_{sr} = \sigma_{sr}/\sqrt{\sigma_{ss}\sigma_{rr}}$. However, simply analyzing these correlations is not sufficient to distinguish between direct interactions between species and associations resulting from indirect links. For example, the abundance of two prey species that share a predator may be correlated simply as a result of fluctuations in predator abundance, even if there are no direct interactions between these species.

Graphical models (Lauritzen 1996) provide a general probabilistic framework in which this distinction can be made. Without going into detail on the subject of graphical models as a whole, we note that, in the case of Gaussian graphical models (GGM), the *precision* matrix $\mathbf{\Omega} = [\omega_{sr}]_{1\leq s,r\leq S} := \mathbf{\Sigma}^{-1}$ is associated with the *partial* correlations between species $\widetilde{\rho}_{sr} = -\omega_{sr}/\sqrt{\omega_{ss}\omega_{rr}}$. In the Gaussian context, partial correlation is, in fact, a conditional correlation, meaning that $\widetilde{\rho}_{sr} = 0$ if and only if the latent variables Z_{ns} and Z_{nr} are independent conditional on all other $\{Z_{nq}\}_{q\neq s,r}$. The ecological interpretation of this property is that nullity of $\widetilde{\rho}_{sr}$ indicates that there is no direct interaction between species j and k.

8.2.4.2. *Inference*

A key aim of inference in ecological networks is thus to distinguish between indirect and direct associations, generally assumed to be few in number. This hypothesis implies that matrix $\mathbf{\Omega}$ will be sparse, that is, contains a majority of zero terms. Chiquet *et al.* (2019) propose a version of the PLN model designed for network inference, which builds on the model described in equations [8.1] and [8.2] by adding a penalty term to the inference step, with the aim of making matrix $\mathbf{\Omega}$ sparse. Put more precisely, the parameters of the model are estimated by maximizing the function

$$J_\lambda(\mathbf{Y};\psi,\theta) := J(\mathbf{Y};\psi,\theta) - \lambda\sum_{s<r}|\omega_{sr}|, \qquad [8.7]$$

where $J(\mathbf{Y};\psi,\theta)$ is the lower bound defined in equation [8.4]. The regularization parameter λ controls the sparsity of the matrix $\mathbf{\Omega}$: the higher the value of λ, the fewer interactions ($\widehat{\omega}_{sr} \neq 0$) will be inferred. This objective function is convex in both ψ and θ, meaning that an efficient gradient descent algorithm can be used. The choice of λ is clearly critical; a value may be determined using a penalized likelihood criterion, of the BIC or eBIC type (Foygel and Drton 2010), or by re-sampling (e.g. see Liu *et al.* 2010).

8.3. Data analysis: marine species

8.3.1. *Description of the data*

The data considered here concern the abundance of marine species (fish, invertebrates, algae, etc.) observed on Anapaca island off the coast of southern California. The island is made up of three islets; for the purposes of our study, we shall consider data for the east (AEI) and middle (AMI) sites. Observations will only be taken into account for years in which the coastline was protected: 1999–2014 (16 years) for AEI and 2003–2014 (12 years) for AMI. For each islet (or site), different observation regions were defined: east (E), center (CEN) and west (W). Finally, each region was divided into zones according to distance from the coast: INNER, MID and OUTER. Four observation protocols were defined, adapted for the species in question, using transects (a form of virtual underwater "corridor") at different depths to observe fish, or quadrats of different dimensions on the sea floor for algae, invertebrates and certain fish species. Thus, not all protocols were systematically used in the transects, and certain species were observed using multiple protocols.

Raw data were aggregated by sample, defined as a unique combination of year $\times$ site $\times$ code $\times$ zone, in order to produce a count table. In each sample, the abundance of each species was defined as the total number of occurrences of the species in question across all transects. The sampling effort (used as the offset in PLN models) was defined as the number of transects in which a protocol permitting the observation of the species in question was implemented.

The count table was then filtered, retaining only (i) samples with a strictly positive observation intensity for at least 80% of species and (ii) species with a mean abundance greater than 1 and a strictly positive sampling effort in at least 80% of the remaining samples. After filtering, 66 species (from a total of 195) and 142 samples (from a total of 169) were retained. This step is necessary to remove null observation intensities, which cannot be processed numerically, and to reduce the proportion of null counts in the table (from 76% down to 44%).

The data obtained from this pre-processing of raw species count data are known as the MariNet data. These data consist of (i) a definition of samples, (ii) the abundance of each species and (iii) the sampling effort for each sample; we can thus construct three matrices, $\mathbf{Y}$, $\mathbf{X}$ and $\mathbf{O}$. Species are identified by a code in the text; these codes are shown in Table 8.1 alongside the scientific and common English name of the species.

	Code	Scientific name	Common name/description
Fish	BFRE	*Brachyistius frenatus*	Kelp surfperch
	EMOR	*Engraulis mordax*	Northern anchovy
	KGB	*Sebastes (atrovirens, carnatus, chrysomelas, caurinus)*	Rockfish
	SJAP	*Scomber japonicus*	Greenback mackerel
	TSYM	*Trachurus symmetricus*	Jack mackerel
Invertebrates	ANTSOL	*Anthopleura sola*	Green anemone
	APLCAL	*Aplysia californica*	California brown
	BARNAC		Barnacle
	CYPSPA	*Cypraea spadicea*	Chestnut cowrie
	DICTYOTALES	*Dictyota spp.* and *Dictyopteris undulata*	
	LOPCHI	*Lophogorgia chilensis*	Red gorgonian
	LYTANAAD	*Lytechinus anamesus*	White urchin, adult > 2.5 cm
	MEGSPP	*Megastrea spp.*	Turban snail
	MEGUND	*Megastrea undosum*	Wavy turban snail
	PANINT	*Panulirus interruptus*	Spiny lobster
	STRFRAAD	*Strongylocentrotus franciscanus*	Red urchin, adult > 2.5cm
	STRPURAD	*Strongylocentrotus purpuratus*	Purple urchin, adult > 2.5cm
Algae	BROWN	*Colpomenia spp.*	Brown algae
	BUSHY	*Gelidium, Pterocladia, Gastroclonium, Gracilaria, Condracanthus canaliculatus*	Red algae with cylindrical branches
	CYSOSMAD	*Cystoseira osmundacea*	Bladder chain, adult diameter > 6cm
	ENCRED		Encrusting non-coralline red algae
	LAMSPP	*Laminaria spp.*	
	MACPYR_HF	*Macrosystis* holdfast	
	MACPYRAD	*Laminaria spp.*	Giant kelp, adult height
	PTECALAD	*Pterygophora californica*	

Table 8.1. *Glossary of species codes, scientific names and common names for species which, according to the model, interact with one of the urchin species, STRFRAAD or STRPURAD*

8.3.2. *Effects due to site and date*

We began by using the Poisson log-normal model including the effect of several covariates, one by one, in order to determine which have the strongest influence. These covariates include site, observation side, zone and period. The period is defined as a group of successive years. Initial analysis did not reveal a strong "year" effect, but when years are grouped into two periods (1999–2001 and after 2001), a notable difference emerges. The cutoff point in 2001 was obtained automatically based on the results of an ascending hierarchical classification (using the Ward criterion) applied to the dataset. The covariates with the strongest effect were identified using the BIC model selection criterion (ICL is better-suited to clustering problems). The BIC was maximized for the model including a site effect and the model including a period effect. The BIC of the other models was lower than that of the covariate-free model.

The site effect appears to result from the presence of certain islet-specific species, as we see from the coefficient values for the two sites in the regression model (see Figure 8.3). These species include LAMSPP (alga), PTECALAD (alga), MEGSPP (seasnail) and SJAP (mackerel) for the AEI islet, and TSYM (mackerel) for the AMI islet.

Figure 8.3. *Poisson log-normal model with the "site" covariate. Representation of the coefficients of each site in the regression model. For a color version of this figure, see www.iste.co.uk/peyrard/ecology.zip*

The period effect may be explained by the fact that, in the first years (1999–2001), data were collected from the AEI site alone; this site only gained protected status in 1999. The first period may reflect a transition phase in the composition and structure of the community of species present on the islet, which then stabilized during the second period.

An estimation of the Poisson log-normal model with the "period" covariate for AEI samples alone shows that the difference between the two periods results from a higher incidence of certain species in period 2 (LAMPPS, MEGSPP, TSYM: see Figure 8.4).

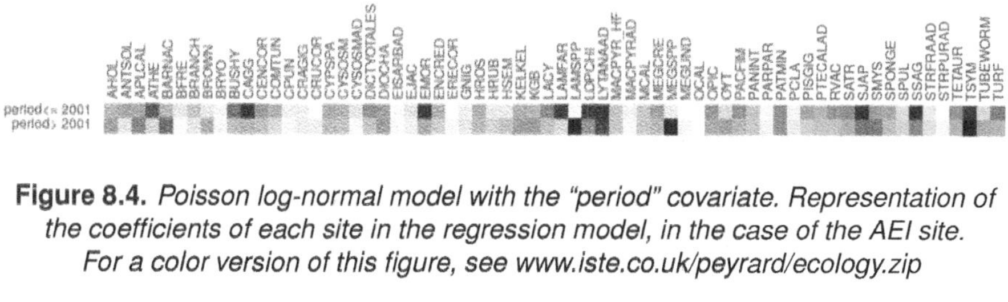

Figure 8.4. *Poisson log-normal model with the "period" covariate. Representation of the coefficients of each site in the regression model, in the case of the AEI site. For a color version of this figure, see www.iste.co.uk/peyrard/ecology.zip*

8.3.3. *Dimension reduction*

Next, we studied the way the model with dimension reduction behaves when no covariate, the site covariate alone (identified as being the most influential during the first stage of our analysis), or all covariates are included. The dimension for each of these three models was chosen using the BIC criterion. The BIC systematically selected 19 dimensions (compared to 66 in the full model), although the variance remained concentrated around the first 5 to 10 axes of the latent space (Figure 8.5, first line).

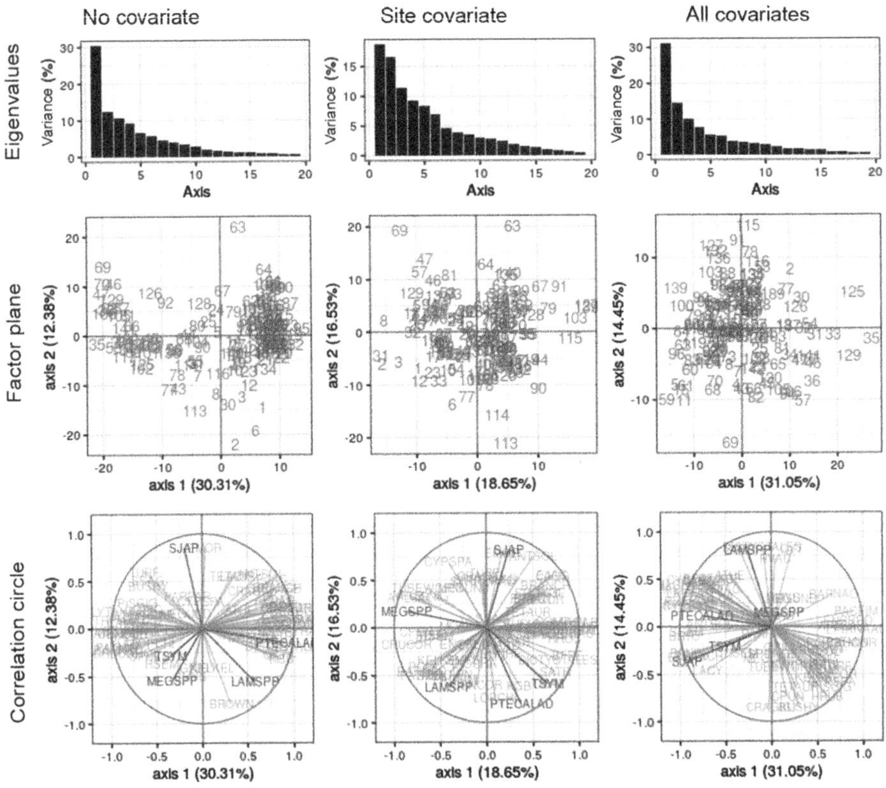

Figure 8.5. *Dimension reduction for (left to right): a model without covariates; a model with the site covariate; and a model with all covariates. First line: plot of the eigenvalues for each dimension. Second line: representation of samples in the first principal plane (blue: AEI; red: AMI). Third line: representation of species on the correlation circle for the first two dimensions (the species in blue are site-specific, as explained above). For a color version of this figure, see www.iste.co.uk/peyrard/ecology.zip*

The addition of the site, followed by the other covariates, reduces their impact on the structure of communities in the latent space (Figure 8.5, second line), allowing us to focus on the residual space. Formally, the site effect is still present in the latent positions $\widetilde{\mathbf{Z}}_n$. In the first model (no covariate), this was integrated into the residual structure $\widehat{\mathbf{T}}\widetilde{\mathbf{m}}_n$ (and thus visible in the graphics), while in the second and third cases, it disappeared from the residual structure (and thus from graphical representations), moving into the corrective term $t(\mathbf{x}_n)\widehat{\mathbf{B}}$ associated with the covariates. This resulted in a better spacing of species around the correlation circle (Figure 8.5, line 3), notably reducing the number with a strong association with axis 1.

If dimension reduction is applied to samples from the AEI islet alone, in a model with the site effect, the period effect identified in our first analysis is once again apparent; there is a clear difference between samples from period 1 and those from period 2 in the first principal plane (Figure 8.6).

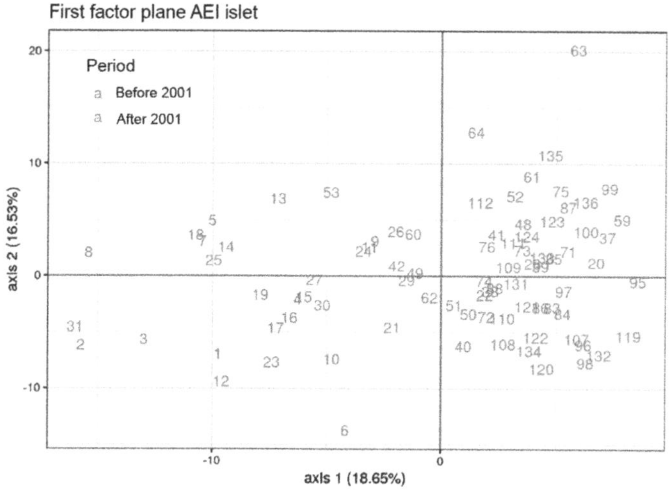

Figure 8.6. *Representation of samples on the AEI islet in the first principal plane. There is a clear difference between samples from period 1 (red) and samples from period 2 (blue). For a color version of this figure, see www.iste.co.uk/peyrard/ecology.zip*

8.3.4. *Inferring ecological interactions*

Our aim now is to identify direct interactions between species using the approach described in section 8.2.4. The network is obtained by "forcing" the precision matrix to contain a large number of zero terms; the proportion of zeros (and thus of connections in the network) is controlled by the parameter λ.

8.3.4.1. *Penalty effect*

Figure 8.7 shows the effect of parameter λ on the density (i.e. the proportion of edges which are present) and fitting of the 16 possible models, obtained by combining the four covariates: year, site, side and zone. As expected, the network density (left) and the model fitting (measured by J_λ, defined in [8.7], right) systematically increase as the regularization parameter λ decreases. The use of a model selection criterion – in this case, BIC – offers the means of correcting this effect and determining an optimal value for λ.

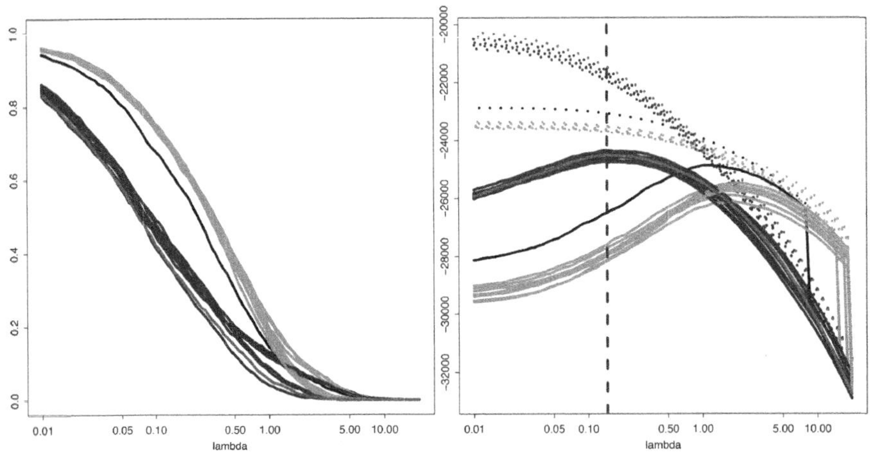

Figure 8.7. *Effect of parameter λ on edge density (left) and model fitting (right). Black = model without covariate, red = full model, green = models without the year effect, blue = models including the year effect. Right: dashed line = lower bound J_λ, solid line = BIC. Vertical dashed line: optimal value of λ for the "year+site" model. For a color version of this figure, see www.iste.co.uk/peyrard/ecology.zip*

Figure 8.7 illustrates two very different behaviors, according to whether the model includes the year effect. Models in which the year is included (blue and red curves) are better fitted and more parsimonious in terms of edges. This observation provides us with confirmation that the year has a major effect on the abundance of different species. Finally, the best BIC score ($-24, 348.52$) is obtained for the model that takes account of both year and site; this optimum is reached for a value of $\lambda = 0.147$.

8.3.4.2. *Robustness of edges*

The choice of the regularization parameter is clearly critical and has a major influence on the final network, notably in terms of density (see Figure 8.7, left). The robustness of the results can be improved by using a re-sampling approach, such as that put forward by Liu *et al.* (2010), which consists of fitting the model to a large number of sub-samples and assigning a selection frequency to each edge. The results of this procedure for the model including both year and site are shown in Figure 8.8.

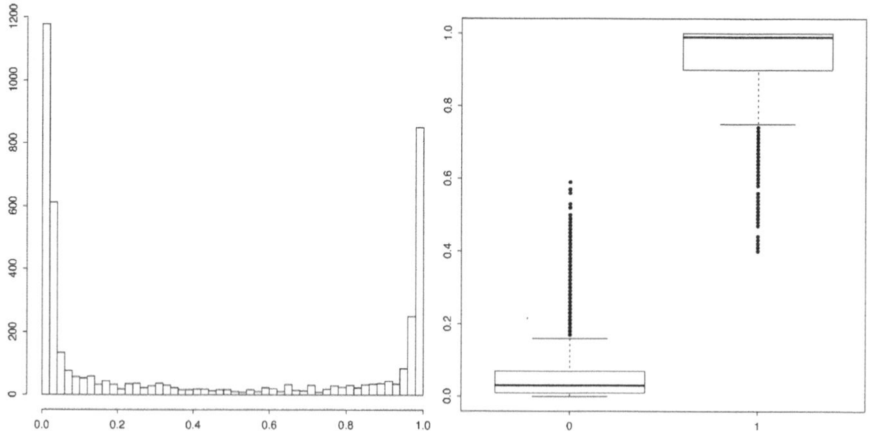

Figure 8.8. *Stability of edges selected for the model including the effects of the year and site covariates. Left: histogram of edge selection frequencies in the sub-samples. Right: Distribution of these frequencies for edges not selected using the BIC ("0") and for edges selected using the same criterion ("1")*

Figure 8.8 (left) shows a clear separation between edges that are almost systematically selected and edges that are almost never selected. On the right side of the figure, we see that in our case the distribution of these frequencies is coherent with the list of edges obtained directly using the BIC. The re-sampling process thus proves the robustness of the edge selection carried out using BIC.

8.3.4.3. *Inferred network*

The inferred network (Figure 8.9) is relatively dense (with a density of 0.4); certain species are linked to many of the other species, corresponding to a complex ecosystem. The complexity of ecological relationships (predation, parasitism, symbiosis, etc.), the non-direction of relationships, the fact that not all species are taken into account and the low number of observations all mean that the network should be analyzed with precaution (Blanchet *et al.* 2020), not as a network of real, direct interactions but rather as a tool to guide reflection in conjunction with present ecological knowledge.

A visualization of the partial correlation matrix (Figure 8.10, with species classified as fish, invertebrates and algae) is helpful for network analysis purposes. This matrix is symmetrical as the relationships are not directed. Note that the region corresponding to interactions between fish and invertebrates or algae contains fewer relationships with weights greater than 0.1 than the rest of the matrix. The frequency is 0.12 for interactions between fish species and 0.04 for interactions between fish x {invertebrates, algae}. Given that fish feed on invertebrates and algae (although some species eat other fish), one would expect to observe stronger feeding relationships between these groups. These interactions may be harder to detect when the relationship is more complex (diversified food supply, habitat, etc.). The relationship between MACPYRAD ("giant kelp", adult) and MACPYR_HF ("giant kelp", holdfast) is also interesting: the weighting (a partial correlation of 0.37) is much higher than any of the others (the next highest weighting, in terms of absolute value, is 0.28). This is due to the fact that both groups are, in fact, the same species at different life stages.

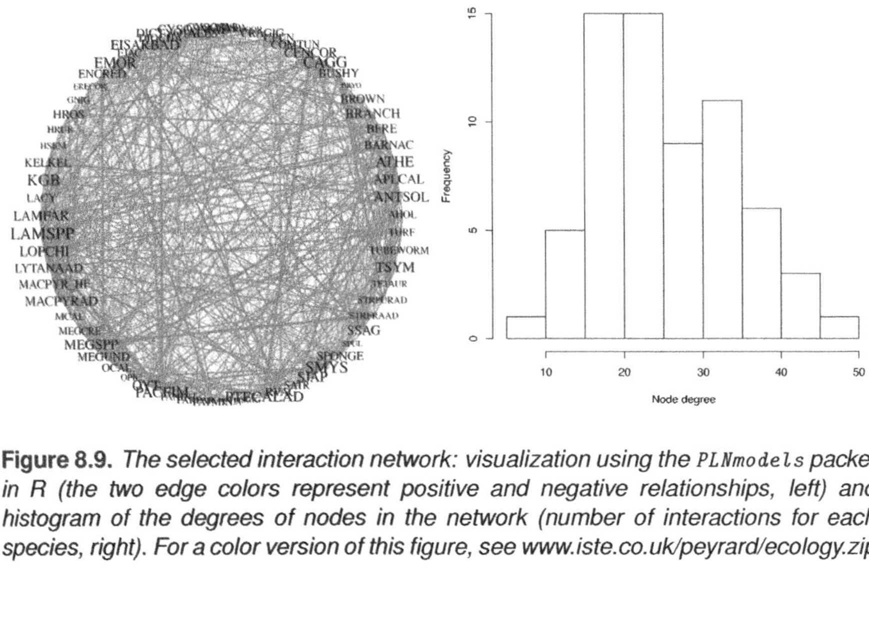

Figure 8.9. *The selected interaction network: visualization using the* `PLNmodels` *packet in R (the two edge colors represent positive and negative relationships, left) and histogram of the degrees of nodes in the network (number of interactions for each species, right). For a color version of this figure, see www.iste.co.uk/peyrard/ecology.zip*

For an ecological study, it may be interesting to analyze the interactions with the highest weight, the species with the most connections, or other species of interest. Here, we have chosen to focus on an invasive species of purple urchin (STRPURAD), which has been colonizing the seabed, to the detriment of an edible red urchin species (STRFRAAD), and the state of the ecosystem as a whole (Woody 2020). Figure 8.11 shows the identified interactions between the two urchin species: the level of interaction is clearly high. The positive sign of the partial correlation associated with the interaction appears hard to explain from an ecological perspective. As STRPURAD is an invasive species, the increase in the size of this population should not, *a priori*, promote an increase in the population of STRFRAAD. However, actions may have been taken to protect STRFRAAD from the damage caused by the increasing STRPURAD population. Each urchin species is in interaction with a similar number of species, with a relatively equal balance of algae, invertebrates and fish. In terms of trophic interactions, this corresponds to the fact that urchins consume algae and are consumed by certain fish species. There are only five species that interact with both of the urchin species in question. This may indicate that the two species have relatively different networks of ecological interactions. Four of these five species are strongly connected to the AEI islet, as identified in our study of covariate effects; they may play a structuring role with respect to the community of this site. Finally, note that several of the species in question are of direct interest for humans: lobsters (PANINT), mackerel (TSYM, SJAP), anchovies (EMOR), kelp surfperch (BFRE) and rockfish (KGB). This may explain why biologists have chosen to study the role of urchins in the ecosystem.

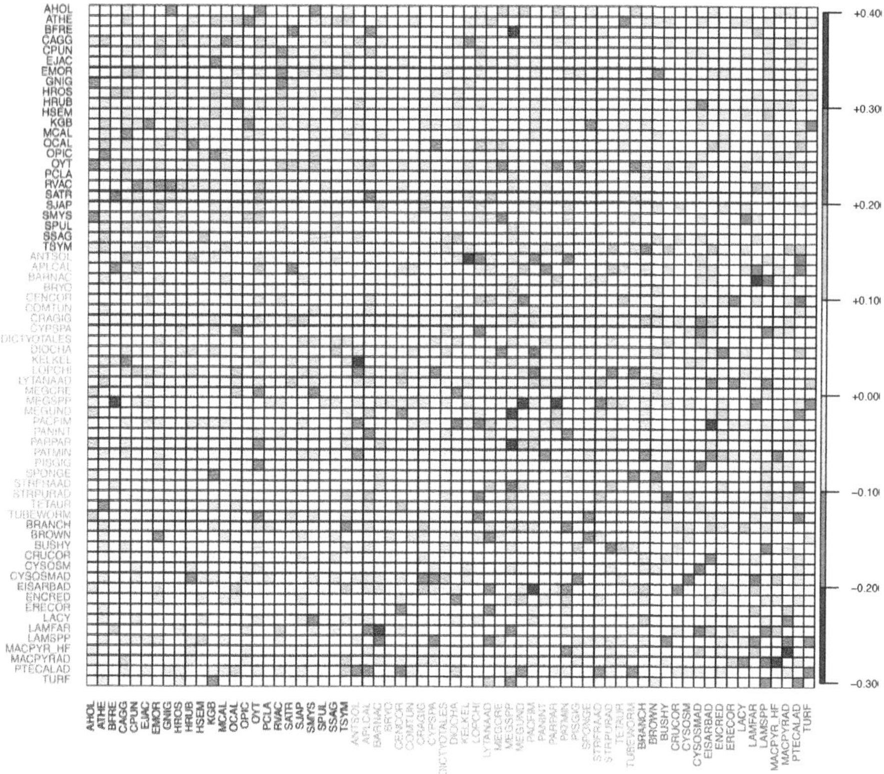

Figure 8.10. *The selected interaction network: visualization of the partial correlation matrix. Species are divided into groups: fish (blue), invertebrates (gray) and algae (red). For a color version of this figure, see www.iste.co.uk/peyrard/ecology.zip*

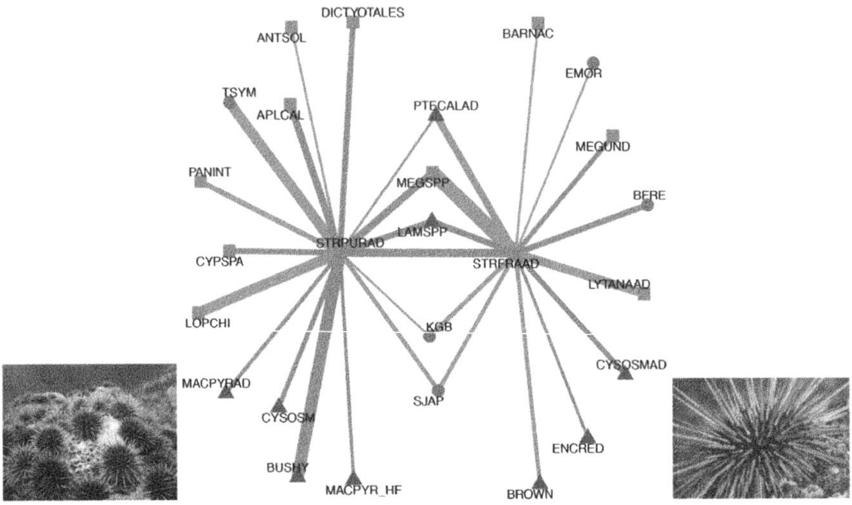

Figure 8.11. *Interactions between two sea urchin species: red (STRFRAAD, Strongylocentrotus franciscanus, left, from `piscoweb.org`) and purple (STRPURAD, Strongylocentrotus purpuratus, right, from source `piscoweb.org`). The round nodes correspond to fish, square to invertebrates and triangle to algae. Red edges represent a positive relationship and light blue edges represent a negative relationship. The thickness of the edges is proportional to the intensity of the relationship. For a color version of this figure, see www.iste.co.uk/peyrard/ecology.zip*

8.4. Discussion

In order to understand how an ecosystem works, we need to understand the interactions that take place between the species making up this ecosystem. In this context, effective statistical analyses rely on a joint modeling of the abundance of all of the species involved. PLN models are flexible and easy to interpret, providing valuable tools for understanding both environmental effects (abiotic interactions) and dependency structures between species (biotic interactions). Like most joint species distribution models (JSDM), the PLN uses a latent layer to model dependencies between species. Unlike other models presented in this book, the latent variables here do not represent a biological reality, but act as an auxiliary element in the modeling process.

Several variants of the PLN model have been presented in this chapter, selected for their ecological relevance; we have described methods for dimension reduction, site or sample comparison and network inference. The ease of interpretation inherent in the PLN facilitates *a posteriori* comparison of the results of the model with external data; for example, the response of a given species to its environment (described using regression coefficients) can be compared with traits or functional groups of species in order to understand the rules that govern this response. All of the variants presented here are implemented in R using the PLNmodels package[4]. The syntax of this package is similar to that of most models in R. Other generalizations are being developed at the time of writing and should be available in the near future. One notable development involves the introduction of a Gaussian mixture model (McLahan and Peel 2000) into the latent part of the model [8.1] in order to structure sites as homogeneous groups.

Note that dimension reduction, as presented in section 8.2.3, is particularly helpful in cases involving a large number of species S. This forms a counterpart to the SCGLR method presented in Chapter 9 for cases involving a high number of covariates D: the aim of SGCLR is to automatically display a smaller number of explanatory components, defined as combinations of original covariates. As we saw in section 8.2.2, the inference method used here relies on a variational approximation, which increases computational efficiency but does not allow us to calculate measures of uncertainty (standard deviation, confidence interval) for parameter estimations, or to define significance tests. Further work is currently ongoing in these areas.

The PLN model offers a means of taking account of differences in sampling effort across sites, samples or species, an essential consideration in avoiding bias and guaranteeing the relevance of results. In many cases, no direct measure of this effort is available, meaning that only estimations (e.g. the total number of observed individuals) can be used. The quality of these estimations has an obvious influence on the quality of the study's results.

8.5. Acknowledgments

The authors wish to thank Jennifer Caselle for providing us with data from the PISCO program and Jennifer Caselle and Laura Dee for valuable discussions concerning the interpretation of these data.

8.6. References

Aitchison, J. and Ho, C. (1989). The multivariate Poisson-log normal distribution. *Biometrika*, 76(4), 643–653.

Blanchet, F.G., Cazelles, K., Gravel, D. (2020). Co-occurrence is not evidence of ecological interactions. *Ecology Letters*, 23(7), 1050–1063 [Online]. Available at: https://onlinelibrary.wiley.com/doi/abs/10.1111/ele.13525.

4 Available to download at cran.r-project.org.

Caselle, J. (2013). A decade of protection, 10 years of change at the channel islands [Online]. Available at: http://www.piscoweb.org/sites/default/files/portfolios/CI_10-Yr_Brochure_web.pdf.

Chiquet, J., Mariadassou, M., Robin, S. (2018). Variational inference for probabilistic Poisson PCA. *The Annals of Applied Statistics*, 12(4), 2674–2698.

Chiquet, J., Mariadassou, M., Robin, S. (2019). A variational Bayesian framework for graphical models. *International Conference on Machine Learning*, Long Beach, CA, USA.

Clark, J.S., Gelfand, A.E., Woodall, C.W., Zhu, K. (2014). More than the sum of the parts: Forest climate response from joint species distribution models. *Ecological Applications*, 24(5), 990–999.

Dempster, A.P., Laird, N.M., Rubin, D.B. (1977). Maximum likelihood from incomplete data via the EM algorithm. *Journal of the Royal Statistical Society, Series B*, 39, 1–38.

Elith, J. and Leathwick, J.R. (2009). Species distribution models: Ecological explanation and prediction across space and time. *Annual Review of Ecology, Evolution, and Systematics*, 40, 677–697.

Foygel, R. and Drton, M. (2010). Extended Bayesian information criteria for Gaussian graphical models. *Advances in Neural Information Processing Systems*, 604–612.

Harris, D.J. (2015). Generating realistic assemblages with a joint species distribution model. *Methods in Ecology and Evolution*, 6(4), 465–473.

Inouye, D.I., Yang, E., Allen, G.I., Ravikumar, P. (2017). A review of multivariate distributions for count data derived from the Poisson distribution. *Wiley Interdisciplinary Reviews: Computational Statistics*, 9(3), e1398.

Johnson, S.G. (2011). The NLopt nonlinear-optimization package.

Lauritzen, S.L. (1996). *Graphical Models*, volume 17. Clarendon Press, Oxford.

Layeghifard, M., Hwang, D.M., Guttman, D.S. (2017). Disentangling interactions in the microbiome: A network perspective. *Trends in Microbiology*, 25(3), 217–228 [Online]. Available at: http://www.sciencedirect.com/science/article/pii/S0966842X16301858.

Liu, H., Roeder, K., Wasserman, L. (2010). Stability approach to regularization selection (stars) for high dimensional graphical models. *NIPS'10, Curran Associates Inc.*, 1432–1440.

McLahan, G. and Peel, D. (2000). *Finite Mixture Models*. John Wiley & Sons, Inc, Brisbane.

Ovaskainen, O., Tikhonov, G., Norberg, A., Blanchet, F.G., Duan, L., Dunson, D., Roslin, T., Abrego, N. (2017). How to make more out of community data? A conceptual framework and its implementation as models and software. *Ecology Letters*, 20(5), 561–576.

Pan, J., Marcoval, M., Bazzini, S., Vallina, M., De Marco, S. (2013). Coastal marine biodiversity: Challenges and threats. In *Marine Ecology in a Changing World*, Arias, A.h., Menendez A.C. (eds.). CRC Press, Boca Raton.

PISCO Research Consortium (2019a). Kelp forest sampling protocols [Online]. Available at: http://www.piscoweb.org/kelp-forest-sampling-protocols.

PISCO Research Consortium (2019b). Partnership for interdisciplinary studies of coastal oceans [Online]. Available at: http://piscoweb.org.

Popovic, G.C., Hui, F.K., Warton, D.I. (2018). A general algorithm for covariance modeling of discrete data. *Journal of Multivariate Analysis*, 165, 86–100.

Schwarz, G. (1978). Estimating the dimension of a model. *Annals of Statistics*, 6(2), 461–464 [Online]. Available at: http://dx.doi.org/10.1214/aos/1176344136.

Svanberg, K. (2002). A class of globally convergent optimization methods based on conservative convex separable approximations. *SIAM Journal on Optimization*, 12(2), 555–573.

Tipping, M.E. and Bishop, C.M. (1999). Probabilistic principal component analysis. *Journal of the Royal Statistical Society: Series B (Statistical Methodology)*, 61(3), 611–622 [Online]. Available at: http://dx.doi.org/10.1111/1467-9868.00196.

Tylianakis, J.M., Laliberté, E., Nielsen, A., Bascompte, J. (2010), Conservation of species interaction networks. *Biological Conservation*, 143(10), 2270–2279 [Online]. Available at: http://www.sciencedirect.com/science/article/pii/S0006320709005126.

Wainwright, M.J. and Jordan, M.I. (2008). Graphical models, exponential families, and variational inference. *Foundations and Trends in Machine Learning*, 1(1–2), 1–305.

Warton, D.I., Blanchet, F.G., O'Hara, R.B., Ovaskainen, O., Taskinen, S., Walker, S.C., Hui, F.K. (2015). So many variables: Joint modeling in community ecology. *Trends in Ecology & Evolution*, 30(12), 766–779.

Woody, T. (2020). California's critical kelp forests are disappearing in a warming world. Can they be saved? [Online]. Available at: https://www.nationalgeographic.com/science/2020/04/california-critical-kelp-forests-disappearing-warming-world-can-they-be-saved/.

Xiao, H., Dee, L.E., Chadès, I., Peyrard, N., Sabbadin, R., Stringer, M., McDonald-Madden, E. (2018). Win-wins for biodiversity and ecosystem service conservation depend on the trophic levels of the species providing services. *Journal of Applied Ecology*, 55(5), 2160–2170 [Online]. Available at: https://besjournals.onlinelibrary.wiley.com/doi/abs/10.1111/1365-2664.13192.

9

Supervised Component-Based Generalized Linear Regression: Method and Extensions

Frédéric MORTIER[1,2], Jocelyn CHAUVET[3,4], Catherine TROTTIER[5,6], Guillaume CORNU[1] and Xavier BRY[6]

[1]CIRAD, UPR Forêts et Sociétés, Montpellier, France.
[2]Forêts et Sociétés, University of Montpellier, CIRAD, France.
[3]INSERM U1219 BPH, University of Bordeaux, France
[4]Centre de recherche de l'ICES, La Roche-sur-Yon, France
[5]Paul-Valéry Montpellier 3 University, France
[6]Institut Montpelliérain Alexander Grothendieck, University of Montpellier, CNRS, France

9.1. Introduction

Changes at a global level have resulted in the modification of ecosystems and their operations. Direct results include species extinctions and changes to the flora, fauna and functional nature of ecosystems, while indirect results include changes to the services which these systems render. Understanding and predicting the impact of these changes in terms of the distribution of species is crucial in order to promote sustainable management strategies. Mathematical and statistical modeling tools are increasingly widely used to quantify and test the effect of changes on biodiversity and ecosystemic services.

Generalized linear models (GLMs; Nelder and Wedderburn 1972) are a classic solution, offering an elegant mathematical framework for understanding a wide variety of real-world situations. These models draw on the exponential family of

Statistical Models for Hidden Variables in Ecology,
coordinated by Nathalie PEYRARD and Olivier GIMENEZ. © ISTE Ltd 2022.

distributions, which includes well-known examples such as the Gaussian, Poisson, Bernoulli and multinomial distributions. A link function connects the expectation of the variable of interest with a set of explanatory variables (McCullagh and Nelder 1989). The choice of this function introduces a high degree of flexibility, meaning that many different ecosystem characteristics can be modeled in this way. Examples include wood density, biomass, carbon, deciduousness, or the abundance and presence or absence of a species. This methodological framework forms one of the bases for species distribution models (SDMs; Guisan and Zimmermann 2000; Elith and Leathwick 2009). The extension of SDMs to multivariate cases is the object of much current study in statistical ecology, as evidenced by the development of joint species distribution models (JSDMs; Warton *et al.* 2015). They are built on multivariate *probit* models (Pollock 2014; Wilkinson *et al.* 2019), in the case of presence–absence data, or on multivariate Poisson log-normal (PLN) models for abundance data (Aïtchison and Ho 1989; Chib and Winkelmann 2001, see Chapters 7 and 8). Recent studies into JSDMs have focused on modeling probabilistic dependency structures between species (Warton *et al.* 2015; Wilkinson *et al.* 2019); few, however, have attempted to model inter-species connections by searching for shared explanatory factors. Research in this area is particularly critical due to the large number of predictors that may be encountered in any given situation, a number which may even exceed the number of observations. Redundancies between predictors can lead to over-adjustment phenomena and/or to the emergence of singularities in the estimation process, resulting in unstable estimations and predictions.

A classic approach to managing instability problems in the univariate case consists in reducing the number of explanatory variables using model selection procedures based on information criteria (Hastie *et al.* 2001). However, these approaches lead to optimization problems, becoming increasingly hard as the dimension (i.e. the number of explanatory variables) increases (Fan and Li 2006). Alternative strategies include regularization by penalization or methods based on component construction.

Penalty-based approaches aim to maximize the likelihood, penalized by a certain function of the regression coefficients vector (Bickel *et al.* 2006). In the case of ridge regression (Hoerl and Kennard 1970), the penalization function is proportional to the norm L_2 of the coefficient vector, while least absolute shrinkage and selection operator (LASSO) regression uses norm L_1(Tibshirani 1996). LASSO regression is particularly suitable in cases where the "true" vector of the regression coefficients is hollow (i.e. contains a large number of zeros). This method is mainly used for variable selection purposes. Unfortunately, it is not an efficient means of selecting groups of variables, and it is not guaranteed to be optimal in high-dimension cases (Fan and Li 2001). A variety of extensions and alternatives have been developed to cope with these problems. One example is the elastic net (Zou and Hastie 2005), which combines norms L_1 and L_2, so as to combine the benefits of both ridge and LASSO regressions; another example is Smoothly Clipped Absolute Deviation

(SCAD; Fan and Li 2001, 2006; Fan and Lv 2010) penalization, which is suitable for use with high dimension.

The second type of regularization method involves replacing the initial explanatory variables with a small number of linear combinations of these variables, called "components", in order to summarize the information contained in the explanatory variables providing the best prediction of responses. This approach is based on the idea that unknown and non-observed components are "better" regressors than the initial variables. In this respect, components may be assimilated to deterministic latent variables. The main advantage of component-based regularization is that the model is easy to interpret, via the decomposition of a linear predictor over a small number of interpretable directions of the explanatory set. The first component-based regression method is principal component regression (PCR; Kendall 1957; Jolliffe 1982), where the response is regressed over the components that reflect the maximum variability of the explanatory variable set. PCR does not take account of the response when constructing components; for this reason, partial least squares regression (PLSR; Wold 1966) is often preferred. In this case, the constructed components maximize their empirical covariance with the response. However, due to the criterion which must be optimized, PLSR only works with quantitative responses.

The first component-based regularization approach in the context of the univariate GLMs (Marx 1996) consists in introducing the PLS mechanism into the iteratively reweighted least squares (IRLS) estimation algorithm. Supervised component-based generalized linear regression (SCGLR; Bry *et al.* 2013) is an extension of the previous method for multivariate cases. As SCGLR is a PLS-type approach, the constructed components are built on strong structures within the explanatory variables, and must also perform well in terms of response prediction. However, the criterion used to construct components in SCGLR is much more flexible than that used in PLSR; among other things, it is possible to specify the type of structure with which components should align in the explanatory sub-space. Both PCR and PLSR may be seen as special instances of SCGLR.

The broad outlines of SCGLR are presented in section 9.2.1. The main aim of this approach is to provide a flexible dimension reduction tool, using component construction, in the context of multivariate GLMs. In section 9.2.2, this method is extended to the case where explanatory variables are divided into thematic groups. A further extension of SCGLR, which gives added flexibility with respect to the independence of observations, is presented in section 9.2.3. In section 9.3, all three methods are applied to the `genus` dataset in the `SCGLR` package, in order to model and predict the abundance of 15 common species in tropical rainforests in the Congo basin.

9.2. Models and methods

9.2.1. *Supervised component-based generalized linear regression*

Let us consider a multivariate model in which a set of S responses (species) $y_{n1}, \ldots, y_{nS}$ is observed for each individual (site) $n \in \{1, \ldots, N\}$. Let each response y_{ns} be the realization of a random variable Y_{ns} the distribution of which belongs to the exponential family. It is important to note that each response vector $\mathbf{Y}_s = (Y_{1s}, \ldots, Y_{Ns})'$ may have its own distribution, meaning that different measurement processes can be taken into account. This is helpful when observing multiple species of trees, for example, as some may be measured using a presence–absence approach while others may be measured in terms of abundance (implying, for example, a Bernoulli and a Poisson distribution, respectively). Let $\mathbf{Y} = \left[\mathbf{y}_1 \,|\, \cdots \,|\, \mathbf{y}_S\right]$ be a matrix of dimension $N \times S$ with the S observed response vectors making up the columns. Let $\boldsymbol{X} = \left[\boldsymbol{x}_1 \,|\, \cdots \,|\, \boldsymbol{x}_J\right]$ define a matrix of dimension $N \times J$, in which $\boldsymbol{x}_j = (x_{1j}, \ldots, x_{Nj})'$ denotes the observation vector of the jth explanatory variable. In this case, J is presumed to be high, leading to redundancies between explanatory variables. Similarly, $\boldsymbol{A} = \left[\boldsymbol{a}_1 \,|\, \cdots \,|\, \boldsymbol{a}_R\right]$ is a matrix of dimension $N \times R$ consisting of R additional explanatory variables with low redundancy, considered *a priori* to be important in predicting $\mathbf{Y}$. Unlike J, we shall presume R to be low. We shall suppose that the variables in $\boldsymbol{X}$ contain only $H < J$ latent dimensions, which are relevant for modeling and predicting $\mathbf{Y}$; the marginal effects of the variables in $\boldsymbol{A}$, on the other hand, must be precisely quantified. We therefore need to identify and interpret the H latent dimensions of $\boldsymbol{X}$ by constructing components, while retaining the variables contained in $\boldsymbol{A}$.

9.2.1.1. *The SCGLR approach*

SCGLR operates within the statistical framework of generalized linear regression, formalized as follows:

$$\begin{aligned} \forall s \in \{1, \ldots, S\}, \quad & \mathbf{Y}_s \sim \mathcal{F}_s(\mathbf{y}_s, \boldsymbol{\eta}_s) \\ \text{with} \quad & \boldsymbol{\eta}_s = \boldsymbol{X}\boldsymbol{\beta}_s + \boldsymbol{A}\boldsymbol{\delta}_s, \end{aligned} \qquad [9.1]$$

where $\mathcal{F}_s$ is a distribution of the exponential family and $\boldsymbol{\eta}_s$ is the linear predictor used to connect the expectation of $\mathbf{Y}_s$ to the explanatory variables via a link function. For example, in the case of presence–absence data, $\mathcal{F}_s$ is a Bernoulli distribution and the link functions used are logit or probit. For count data, $\mathcal{F}_s$ is a Poisson distribution and the link function log is generally used. In SCGLR, for cases with just one component to identify, the linear predictor is written as:

$$\boldsymbol{\eta}_s = \boldsymbol{X}\boldsymbol{u}\gamma_s + \boldsymbol{A}\boldsymbol{\delta}_s, \qquad [9.2]$$

where $\boldsymbol{u}$ is a unit vector (i.e. $\|\boldsymbol{u}\| = 1$) of length J to determine. In [9.2], $\boldsymbol{f} = \boldsymbol{X}\boldsymbol{u}$ denotes the component to identify, γ_s the associated regression parameter and $\boldsymbol{\delta}_s$ the vector of the parameters related to the additional explanatory variables. It is important to note here that the component $\boldsymbol{f}$ is *common to all of the responses* $\mathbf{Y}$. It acts as a deterministic latent variable, taking account of the structural dependency between responses. Parameters γ_s and $\boldsymbol{\delta}_s$, on the other hand, are specific to the response $\mathbf{y}_s$ and reflect the effect of the component and of the additional explanatory variables for the sth response. This way of writing is only slightly different from a simple regression. It also makes computation more complicated, as estimation can no longer be carried out directly: the product $\boldsymbol{u}\gamma_s$ results in a nonlinearity. However, there are at least two benefits to this approach. First, it offers a natural means of reducing the dimension of the problem, increasing the stability of estimations. Second, the construction of $\boldsymbol{u}$ is flexible and can be adjusted depending on which structures are considered to be most relevant among the explanatory variables. This flexibility will be discussed in greater detail in the next paragraph.

9.2.1.2. *Notion of structural relevance*

Structural relevance can be measured in different ways, depending on the type of direction the component is intended to reflect (Bry and Verron 2015; Bry *et al.* 2020). In this case, we have chosen to use an approach called Variable Powered Inertia (VPI) (Bry and Verron 2015). If $\boldsymbol{X}$ contains J standardized numerical variables, then this measure, denoted as ϕ_ℓ, takes the form of a weighted generalized mean of the squared correlations between explanatory variables $\boldsymbol{x}_j$ and the component $\boldsymbol{f}$. It is defined by

$$\phi_\ell(\boldsymbol{u}) = \left\{\sum_{j=1}^{J} \omega_j \left[\text{cor}^2(\boldsymbol{f}, \boldsymbol{x}_j)\right]^\ell\right\}^{1/\ell}, \qquad [9.3]$$

where ω_j represents the weight of the jth explanatory variable ($\omega_j = 1/J$ by default) and where the scalar $\ell \geqslant 1$ measures the locality (or narrowness) of the desired bundles, that is, groups of highly inter-correlated variables. Figure 9.1 illustrates the role of parameter ℓ, showing the behavior of $[\phi_\ell(\boldsymbol{u})]^\ell$ for different values of ℓ in an elementary case with four coplanar explanatory variables $\boldsymbol{x}_1, \boldsymbol{x}_2, \boldsymbol{x}_3, \boldsymbol{x}_4$. The higher the value of ℓ, the more local, and numerous, the bundles will be.

Note that the VPI measure can be used for qualitative explanatory variables, and is one specific instance of a highly flexible family of structural relevance measures. For more details, see Bry and Verron (2015).

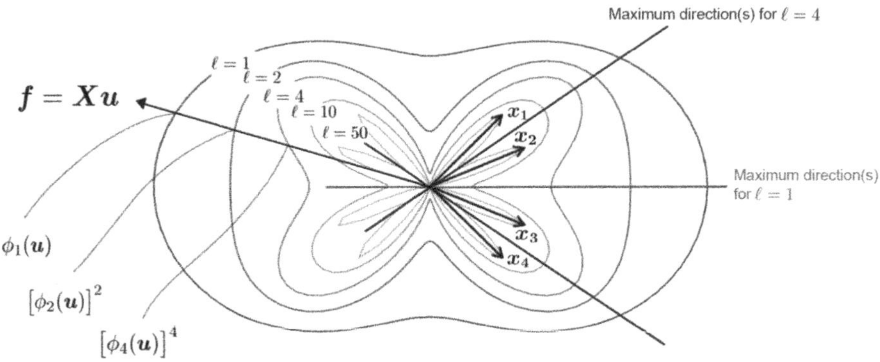

Figure 9.1. *Polar representation of VPI (equation 9.3) by values of ℓ for four coplanar explanatory variables x_1, x_2, x_3, x_4. Vector u is identified with the complex number $e^{i\theta}$, where $\theta \in [0, 2\pi[$. The curves $z_\ell(\theta) = \left[\phi_\ell\left(e^{i\theta}\right)\right]^\ell e^{i\theta}$ are shown for $\ell \in \{1, 2, 4, 10, 50\}$, such that the intersection between the curve z_ℓ and the component $f = Xu$ has a radius equal to $\left[\phi_\ell\left(e^{i\theta}\right)\right]^\ell$. The red line shows the only maximum direction for $\ell = 1$, which is the first principal component. The four variables are then considered as a single bundle. The blue lines show the two maximum directions for $\ell = 4$, and in this case, the variables are seen as two bundles containing two variables each. Finally, when $\ell = 50$, each variable is considered as a bundle in its own right. For a color version of this figure, see www.iste.co.uk/peyrard/ecology.zip*

9.2.1.3. *Estimation method for a single-component model*

Taking account of only relevant structures in X when constructing u offers no guarantee that the vector obtained in this manner will provide the best prediction of responses. The SCGLR approach aims to identify the vector u which gives the best possible trade-off between structural relevance ϕ_ℓ and the goodness of fit of the generalized linear model. The problem may thus be expressed as a maximization of the criterion

$$C(u, \gamma, \delta) = \lambda \log\left[\phi_\ell(u)\right] + (1 - \lambda) \log\left[\mathcal{L}(\mathbf{Y}, u, \gamma, \delta)\right] \qquad [9.4]$$

under the constraint $\|u\| = 1$, where $\mathcal{L}(\mathbf{Y}, u, \gamma, \delta)$ is the likelihood of the generalized linear model and thus provides information concerning fit quality. Parameter λ is used to weight the influence of each term in the overall criterion to maximize. If $\lambda = 0$, the procedure is simply a maximization of the log-likelihood. If $\lambda = 1$, by contrast, the algorithm will seek the direction which maximizes ϕ_ℓ. In [9.4], the log-likelihood of the model is written as:

$$\sum_{s=1}^{S} \log\left[\mathcal{L}_s(\mathbf{y}_s, u, \gamma_s, \delta_s)\right],$$

where $\mathcal{L}_s(\mathbf{y}_s, \boldsymbol{u}, \gamma_s, \boldsymbol{\delta}_s)$ is the likelihood of the model associated with the response $\mathbf{y}_s$. Again, this way of writing highlights the facts that (i) distributions may differ from one response variable to the next, (ii) the relevant direction $\boldsymbol{u}$ is shared by all of the response variables, (iii) the effect of components is variable specific and (iv) the $\mathbf{Y}_s$ are presumed to be independent conditional on $\boldsymbol{X}$.

The [9.4] criterion is maximized using an iterative approach. Given vector $\boldsymbol{u}$, the regression parameters γ_s and $\boldsymbol{\delta}_s$ are estimated using the Fisher scoring algorithm (FSA; Lange 2012), replacing $\boldsymbol{X}$ with $\boldsymbol{X}\boldsymbol{u}$. Given the regression parameters, the construction of vector $\boldsymbol{u}$ is obtained using the projected iterated normalized gradient (PING; Bry *et al.* 2020). This algorithm, used to maximize any given function on the unit sphere, is described in detail in Chauvet (2018).

9.2.1.4. *Construction of higher rank components*

The case considered thus far involves a single component, $\boldsymbol{f} = \boldsymbol{X}\boldsymbol{u}$. In practice, however, a single direction is often not sufficient to correctly predict all responses, raising the need for other predictive components. In order to avoid redundancy phenomena, an approach ensuring that additional components are orthogonal to existing components will be used. Let $\boldsymbol{F}_h = \left[\boldsymbol{f}_1 | \cdots | \boldsymbol{f}_h\right]$ be the matrix containing the first h constructed components and $\widetilde{\boldsymbol{A}}_h = \left[\boldsymbol{A} | \boldsymbol{F}_h\right]$ the matrix of additional variables incremented by $\boldsymbol{F}_h$. Component $\boldsymbol{f}_{h+1} = \boldsymbol{X}\boldsymbol{u}_{h+1}$ is then obtained by maximizing the criterion

$$C(\boldsymbol{u}, \gamma, \widetilde{\boldsymbol{\delta}}_h) = \lambda \log\left[\phi_\ell(\boldsymbol{u})\right] + (1-\lambda)\log\left[\mathcal{L}(\mathbf{Y}, \boldsymbol{u}, \gamma, \widetilde{\boldsymbol{\delta}}_h)\right],$$

where $\widetilde{\boldsymbol{\delta}}_h$ are the parameters associated with the matrix $\widetilde{\boldsymbol{A}}_h$, under the constraints $\|\boldsymbol{u}\| = 1$ and $\boldsymbol{f}_{h+1} \perp \boldsymbol{F}_h$. The final choice of the number of components is made via cross-validation. For details, see Bry *et al.* (2013, 2020); Chauvet (2018).

9.2.2. *Thematic supervised component-based generalized linear regression (THEME-SCGLR)*

Let us return to the case described in section 9.2.1, but this time, let the matrix $\boldsymbol{X}$ be decomposed *a priori* into T conceptually homogeneous thematic groups:

$$\boldsymbol{X} = \left[\boldsymbol{X}_1 | \cdots | \boldsymbol{X}_T\right].$$

For example, if we wish to model and predict the abundance of tree species, two distinct groups of explanatory variables may be used: one with variables related to environmental conditions (pluviometry, geology, etc.) and one corresponding to the photosynthetic characteristics of the population (which may be summarized using EVI, MIR and NIR indices, obtained via teledetection). This distinction is helpful in distinguishing between the effects of different types (thematic groups) of explanatory variables on responses.

9.2.2.1. *One component per thematic group*

Each thematic group $\boldsymbol{X}_t$ $(t = 1, \ldots, T)$ is presumed to contain a large number J_t of variables that must be synthesized using a small number $H_t < J_t$ of relevant latent dimensions. As in the case of SCGLR, a generalized linear regression (described by 9.1) is implemented, using linear predictors that correspond to a decomposition of $\boldsymbol{X}$ into thematic groups. In cases where only one component is constructed for each thematic group, these are written as:

$$\eta_s = \sum_{t=1}^{T} \boldsymbol{X}_t \boldsymbol{u}_t \gamma_{ts} + \boldsymbol{A}\boldsymbol{\delta}_s, \quad s = 1, \ldots, S,$$

where $\boldsymbol{f}_t = \boldsymbol{X}_t \boldsymbol{u}_t$ denotes the component associated with thematic group t and γ_{ts} is the regression parameter of the response s on this component. The THEME-SCGLR method then consists in constructing vectors $\boldsymbol{u}_1, \ldots, \boldsymbol{u}_T$, which maximize a trade-off between the product of the structural relevances in the groups and the overall goodness of fit of the model. This may be expressed as a maximization of the criterion

$$C(\boldsymbol{u}_1, \ldots, \boldsymbol{u}_T, \gamma_1, \ldots, \gamma_T, \boldsymbol{\delta}) = \lambda \sum_{t=1}^{T} \log\left[\phi_\ell(\boldsymbol{u}_t)\right] + (1-\lambda)\log\left[\mathcal{L}(\mathbf{Y}, \boldsymbol{u}_1, \ldots, \boldsymbol{u}_T, \gamma_1, \ldots, \gamma_T, \boldsymbol{\delta})\right] \quad [9.5]$$

under the constraints $\|\boldsymbol{u}_t\| = 1$ $(t = 1, \ldots, T)$, where $\mathcal{L}$ denotes the likelihood of the generalized linear model and $\gamma_t = (\gamma_{t1}, \ldots, \gamma_{tS})'$ corresponds to the vector of the regression coefficients associated with the thematic group t. Criterion 9.5 can be maximized iteratively, applying the SCGLR approach described in section 9.2.1 sequentially for each thematic group. For a given thematic group t, the criterion to maximize in iteration i is written as:

$$C'\left(\boldsymbol{u}_t^{[i]}, \gamma_t^{[i]}, \widetilde{\boldsymbol{\delta}}^{[i]}\right) = \lambda \log\left[\phi_\ell\left(\boldsymbol{u}_t^{[i]}\right)\right] + (1-\lambda)\log\left[\mathcal{L}\left(\mathbf{Y}, \boldsymbol{u}_t^{[i]}, \gamma_t^{[i]}, \widetilde{\boldsymbol{\delta}}^{[i]}\right)\right],$$

where $\widetilde{\boldsymbol{\delta}}^{[i]}$ is the vector of parameters associated with the matrix $\widetilde{\boldsymbol{A}}^{[i]}$ made up of additional variables incremented by the components in their current state. More precisely, over the course of successive updates from the 1st to the Tth group, for iteration i and group t, the matrix

$$\widetilde{\boldsymbol{A}}^{[i]} = \left[\boldsymbol{A} \,|\, \boldsymbol{f}_1^{[i]} \,|\, \cdots \,|\, \boldsymbol{f}_{t-1}^{[i]} \,|\, \boldsymbol{f}_{t+1}^{[i-1]} \,|\, \cdots \,|\, \boldsymbol{f}_T^{[i-1]}\right]$$

is used.

9.2.2.2. *Multiple components per thematic group*

As we saw in section 9.2.1, and even with several thematic groups, the use of just one component per group may not be sufficient to fully reflect the complexity of observed processes. The search for successive components in groups is based on the same strategy used in SCGLR, with an added constraint of orthogonality between components of the same group, and with iterative modifications to the matrix of additional explanatory variables. However, in the case of multiple thematic groups, the use of cross-validation to select an optimal number of components is more difficult. In order to be exhaustive, the selection process would need to include comparisons of all models, crossing all possible numbers of components in each group. The cost of this combinatorial problem rapidly becomes unfeasibly high. The proposed alternative in this case is the backward ("pruning") method. Starting with a large number of components in each thematic group, the least informative components are successively removed. It is important to note that this approach remains dependent on cross-validation, and that the parameters associated with components are updated each time a component is "pruned".

9.2.3. *Mixed SCGLR*

9.2.3.1. *Beyond independence*

Both SCGLR (section 9.2.1) and its thematic variant THEME-SCGLR (section 9.2.2) rely on a strong hypothesis, that of independent observations. However, in certain real-world situations, observations are not independent, but structured into a number of distinct groups (or clusters), in which observations are *a priori* dependent. The need for increased flexibility with respect to the independence of observations led to the development of the mixed-SCGLR method (Chauvet 2018; Chauvet *et al.* 2019), in which the dependency of observations within each group is modeled by a random effect. This method enables a distinction to be made between the part of variability due to fixed effects and that resulting from the dependency structure of the observations. It extends the SCGLR approach to multivariate generalized linear mixed models (GLMMs; Bolker *et al.* 2009) and offers the possibility of finer interpretation of models.

9.2.3.2. *Presentation of the model*

Once again, let us take the same situation as in section 9.2.1, with the exception that the N observations are no longer presumed independent, but form G distinct groups within which observations are *a priori* dependent. The effect of this dependency structure is presumed different for each response. Thus, for each response $\mathbf{y}_s$, a random effect $\boldsymbol{\xi}_s$ with G levels is used to model the dependency of observations in each group. These random effects are taken to be independent and normally distributed:

$$\forall s \in \{1, \ldots, S\}, \ \boldsymbol{\xi}_s \overset{\text{ind.}}{\sim} \mathcal{N}_G(\mathbf{0}, \boldsymbol{D}_s), \qquad [9.6]$$

where $\boldsymbol{D}_s = \sigma_s^2 \boldsymbol{I}_G$, with σ_s^2 the variance of the random "group" effect associated with the response $\mathbf{y}_s$. Conditional on $\boldsymbol{\xi}_s$, the response vector $\mathbf{y}_s$ is considered to be the realization of a random vector $\mathbf{Y}_s$ with a distribution $\mathcal{F}_s$ belonging to the exponential family. We thus have

$$\forall s \in \{1, \ldots, S\}, \quad \mathbf{Y}_s \mid \boldsymbol{\xi}_s \overset{\text{ind.}}{\sim} \mathcal{F}_s\left(\mathbf{y}_s, \boldsymbol{\eta}_s^{\xi}\right), \qquad [9.7]$$

where the linear predictor $\boldsymbol{\eta}_s^{\xi}$, in the case of a single component model, is written as

$$\boldsymbol{\eta}_s^{\xi} = (\boldsymbol{X}\boldsymbol{u})\,\gamma_s + \boldsymbol{A}\boldsymbol{\delta}_s + \mathbf{Z}\boldsymbol{\xi}_s. \qquad [9.8]$$

In expression [9.8], the matrix $\mathbf{Z}$ of dimension $N \times G$ is the design matrix for random effects, in which an element $z_{n,g}$ is equal to 1 if the observation n belongs to group g and 0 otherwise. The way the linear predictor [9.8] is written clearly shows the way in which different dependency structures within the data can be taken into account. Component $\boldsymbol{f} = \boldsymbol{X}\boldsymbol{u}$, which is common to all responses, is still interpreted as a deterministic latent variable, showing a structural dependency between responses. The random effects $\boldsymbol{\xi}_1, \ldots, \boldsymbol{\xi}_S$ are different, as they model the dependency of observations within each group. They may thus be interpreted as stochastic latent variables, capturing the portion of data variability, which can be ascribed to the presence of grouped data. Unlike the component, the random effects are response dependent, as the effect of the inter-observation dependency structure may differ between responses.

9.2.3.3. *Estimation method*

An adaptation of Schall's algorithm (Schall 1991) can be used to estimate the model described by [9.6]–[9.8]. This is an iterative procedure, alternating between linearization of the model conditional on the random effects and estimation of parameters, using methods based on mixed linear models.

1) *Linearization of the model*: for each $s \in \{1, \ldots, S\}$, this step consists in defining the linearized model

$$\mathcal{M}_s^{\xi} : \left| \begin{array}{l} \boldsymbol{z}_s^{\xi} = (\boldsymbol{X}\boldsymbol{u})\,\gamma_s + \boldsymbol{A}\boldsymbol{\delta}_s + \mathbf{Z}\boldsymbol{\xi}_s + \boldsymbol{e}_s, \\ \text{with: } \mathbb{E}\left(\boldsymbol{e}_s \mid \boldsymbol{\xi}_s\right) = \mathbf{0} \text{ and } \mathbb{V}\left(\boldsymbol{e}_s \mid \boldsymbol{\xi}_s\right) = \mathbf{W}_s^{\xi}, \end{array} \right.$$

where $\boldsymbol{z}_s^{\xi}$ and $\mathbf{W}_s^{\xi}$ respectively denote the pseudo-response and the variance–covariance matrix, as defined by Schall.

2) *Construction of the component and estimation of parameters*: just as in the SCGLR approach (section 9.2.1), the component $\boldsymbol{f} = \boldsymbol{X}\boldsymbol{u}$ is constructed by

maximizing a trade-off between structural relevance and goodness of fit. In this case, the trade-off is written as

$$C(\boldsymbol{u}, \gamma, \boldsymbol{\delta}) = \lambda \log\left[\phi_\ell(\boldsymbol{u})\right] + (1-\lambda) \sum_{k=1}^{q} \log\left[\mathcal{L}_s\left(\boldsymbol{z}_s^{\xi}, \boldsymbol{u}, \gamma_s, \boldsymbol{\delta}_s \,|\, \boldsymbol{\xi}_s\right)\right],$$

where the structural relevance ϕ_ℓ is given by [9.3] and where $\mathcal{L}_s\left(\boldsymbol{z}_s^{\xi}, \boldsymbol{u}, \gamma_s, \boldsymbol{\delta}_s \,|\, \boldsymbol{\xi}_s\right)$ denotes the likelihood of the model $\mathcal{M}_s^{\xi}$. Once component $\boldsymbol{f}$ has been constructed, Schall's method (Schall 1991) is applied using the linear predictors given by [9.8]. The new values of the regression parameters γ_s and $\boldsymbol{\delta}_s$, predictions $\boldsymbol{\xi}_s$ and variance parameters σ_s^2 are obtained using Henderson's method (Henderson 1975).

Steps 1 and 2 are repeated until component $\boldsymbol{f}$ and estimates of $\gamma_s, \boldsymbol{\delta}_s, \sigma_s^2$ stabilize for all $s \in \{1, \ldots, S\}$. The mixed-SCGLR method can also be extended to find $H \geqslant 2$ components by adding extra orthogonality constraints. Details of this extension are presented by Chauvet (2018) and Chauvet *et al.* (2019).

9.3. Case study: predicting the abundance of 15 common tree species in the forests of Central Africa

In this section, we shall illustrate the SCGLR approach using the *genus* dataset (taken from the *SCGLR* package, available from https://github.com/SCnext), which describes the abundance of 15 common tree species in the Congo Basin and 46 geo-referenced explanatory variables. Each of the 1,000 observed individuals in the dataset is a 5 km by 5 km plot, and each plot is the result of an aggregation of a varying number of sub-plots of 0.5 ha. For each plot, there are 21 variables describing the physical characteristics of the environment: altitude, pluviometry, hydric condition of the soil and geology. Geology is a categorical variable with five levels. Vegetation is characterized by 25 indices of photosynthetic activity, obtained by teledetection (using `EVI` (Enhanced Vegetation Index), `MIR` (Middle InfraRed) and `NIR` (Near-InfraRed)). As the sampled surface is not the same from one parcel to the next, surface will be used as a scale parameter (offset) in the model. Finally, spatial coordinates (longitude, latitude) and information related to the forestry concession is also included in the dataset. For reasons of confidentiality, the tree species are denoted as gen_s, $s = 1, \ldots, 15$.

9.3.1. *The SCGLR method: a direct approach*

As a starting point, we shall consider that the matrix $\boldsymbol{X}$ contains all of the explanatory variables with the exception of geology, used as an additional explanatory variable. The abundance data are considered to follow a Poisson distribution, and the canonical link function (log) is used. The choice of the number of components and the tuning parameters (λ and ℓ) are obtained by five-block cross-validation. The maximum number of components is fixed at 10. Figure 9.2 shows the geometric mean of the mean errors. We see that the optimal choice corresponds to $H = 7, \lambda = 0.15$ and $\ell = 4$. In this example, the differences resulting from the use of different values of λ and ℓ are not particularly strong. However, we know from experience that these parameters can have a significant influence on the selection of the number of components, and thus on their interpretation. Furthermore, note that, in this case, the component-free model ($H = 0$) only contains the additional variable (geology).

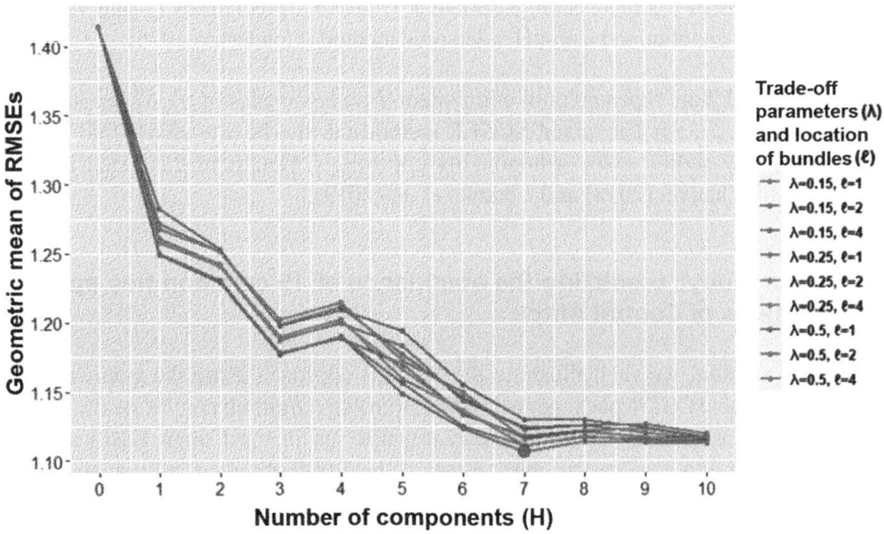

Figure 9.2. *Geometric mean of the square mean quadratic prediction errors over responses as a function of the number of components and for different values of λ and ℓ. The red dot corresponds to the optimal configuration ($H = 7$, $\lambda = 0.15$ and $\ell = 4$). For a color version of this figure, see www.iste.co.uk/peyrard/ecology.zip*

Using the same principle as ACP, Figure 9.3 shows the correlation circles over the planes spanned by the first three components. Variables and linear predictors are only shown for species that are well represented in these planes (correlations greater than 0.8). The first component accounts for 34% of the inertia of the explanatory variables $\boldsymbol{X}$, and is principally structured by the opposition between two contrasting climatic zones: the north, with low rainfall during the dry season (unimodal pluviometry), and the south, where rainfall is more regular over the course of the year. The second axis differentiates between forests with high photosynthetic activity during the dry season and forests with high photosynthetic activity during the rainy season. Axis 3 corresponds to the opposition between highly or weakly partitioned landscapes, structured by the altitude variable. Graphic representations of the correlation circles make the components easy to interpret. Representing linear predictors on factorial planes (red arrows in Figures 9.3a, 9.3b and 9.3c) also assists in understanding the links between environmental characteristics and the abundance of species.

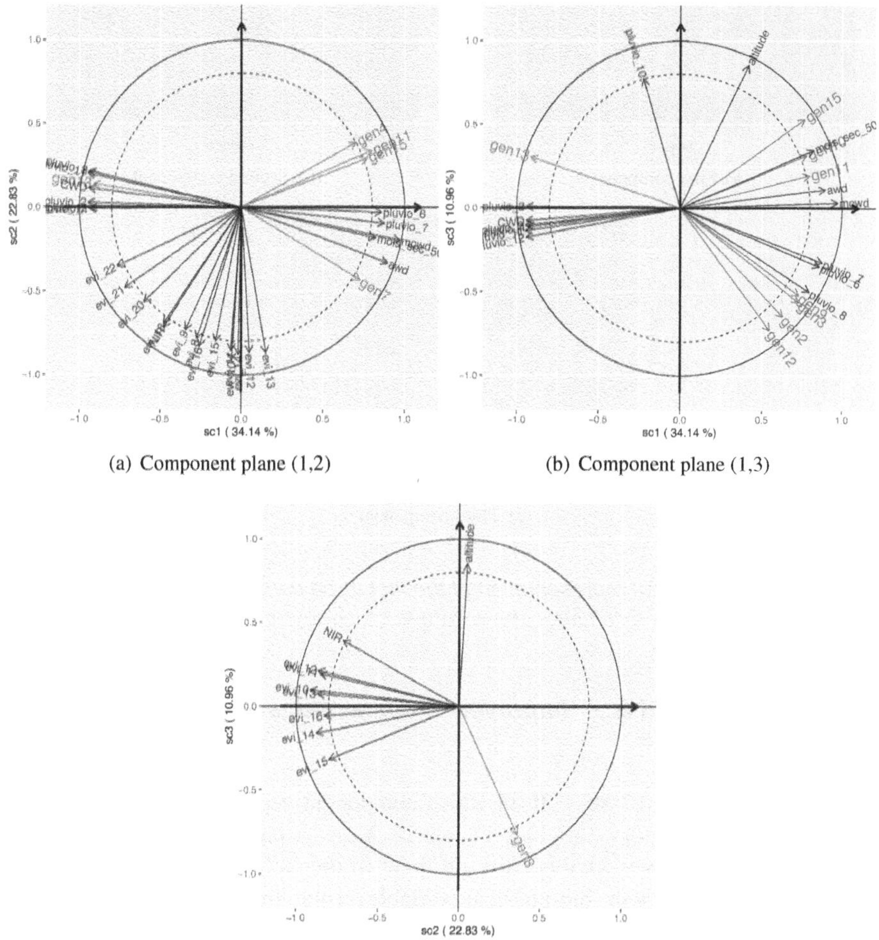

(a) Component plane (1,2)

(b) Component plane (1,3)

(c) Component plane (2,3)

Figure 9.3. *Circles of correlations resulting from the first three components. For a color version of this figure, see www.iste.co.uk/peyrard/ecology.zip*

Finally, components can also be visualized in space by representing components as a function of spatial coordinates; this makes it easier to understand how bundles are structured across the space (see Figure 9.4).

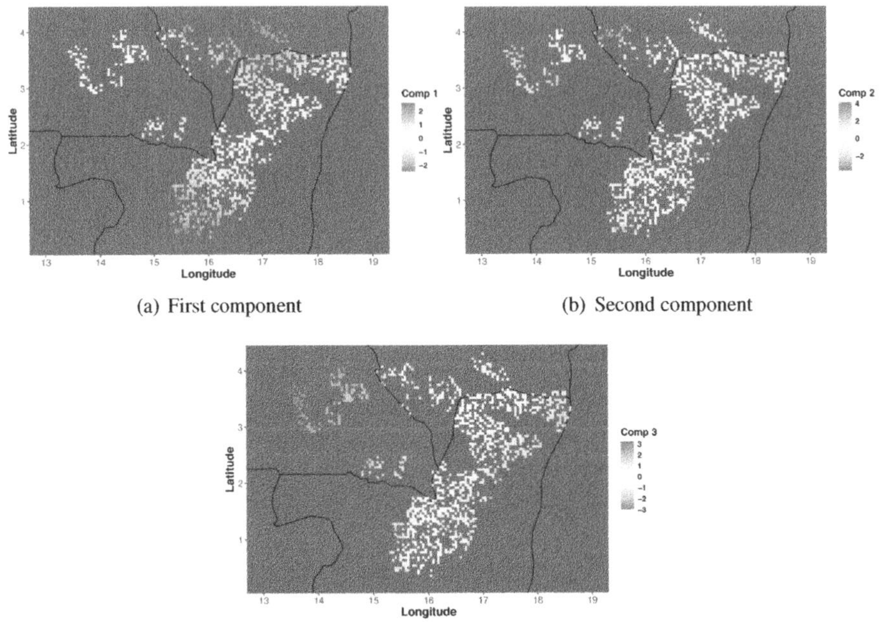

(a) First component (b) Second component

(c) Third component

Figure 9.4. *Spatial representation of the first three components. For a color version of this figure, see www.iste.co.uk/peyrard/ecology.zip*

9.3.2. *THEME-SCGLR: improved characterization of predictive components*

The initial version of SCGLR, in which component vectors are calculated based on all of the explanatory variables in matrix $\boldsymbol{X}$, was presented in section 9.3.1. However, these variables may be very different in thematic terms: in our example, matrix $\boldsymbol{X}$ includes both bio-physical variables (pluviometry, temperature) and variables that characterize photosynthetic activity (`EVI`). The variables represent distinct biological phenomena, and the construction of components combining the two realities can result in interpretation problems. For example, computing the squared correlations (ρ^2) between the first component and the explanatory variables, we see that certain hydric conditions present a very clear correlation to this component (Figure 9.3a). Nevertheless, variables `EVI_21` and `EVI_22` also play a significant role in the definition of this component ($\rho^2 = 0.52$ and 0.58, respectively). The first extension to SCGLR, presented in section 9.2.2, allows components to be constructed within thematic sub-sets of explanatory variables, highlighting the respective role of each theme in species prediction. Here, we shall consider two sub-sets: one containing variables related to the bio-physical environment and a second containing variables that characterize the photosynthetic activity of the population (EVI). The additional variables remain unchanged. The “pruning” approach developed for the THEME-SCGLR method allows us to estimate an “optimal” path between bio-physical variables and those obtained from satellite imaging (see Figure 9.5).

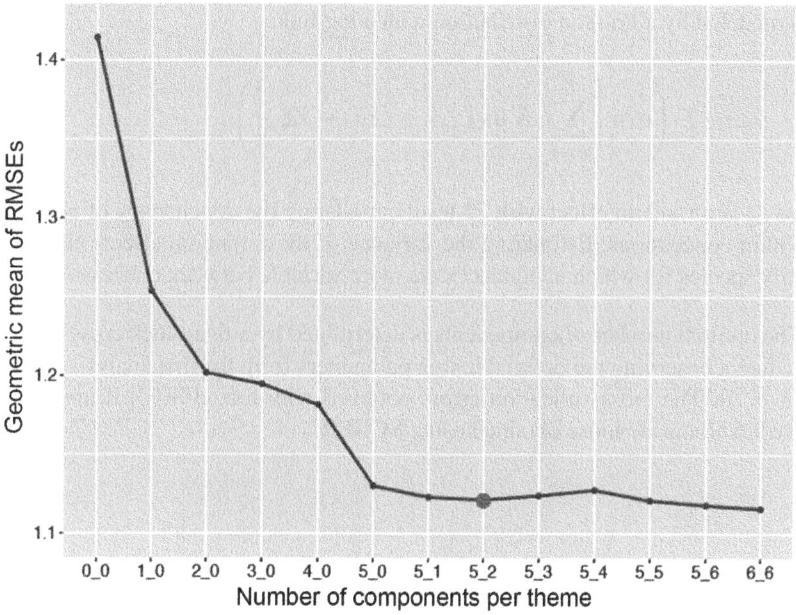

Figure 9.5. *Geometric mean of the square roots of the mean quadratic prediction errors of responses, as a function of the number of components in each theme. On the x axis, m_n corresponds to a model with m components for theme 1 (bio-physical environment) and n components for theme 2 (photosynthetic activity). The optimal model, shown by the red dot, contains five components for theme 1 and two components for theme 2. For a color version of this figure, see www.iste.co.uk/peyrard/ecology.zip*

The optimal number of components here is the same as in our first analysis (7). However, where the second component, calculated on the basis of all explanatory variables, was seen to be strongly linked to photosynthetic activity (EVI variables, see Figure 9.3), the two-group approach shows that these variables are actually of secondary importance compared to bio-physical characteristics. Figure 9.5 should be read from right to left. The model is initially fitted for six components per thematic group, then those components that are least useful for predictive purposes are progressively removed (backwards method); these least useful components mostly concern photosynthetic activity.

9.3.3. *Mixed-SCGLR: taking account of the concession effect*

The next stage is to take account of the fact that the individual plots in the data set belong to 22 different forestry concessions. Measures made from within the same concession will now be considered to be inter-dependent. However, this dependency relationship is expressed differently depending on the species. Each abundance vector $\mathbf{y}_s$ is modeled by a Poisson distribution with a log link:

$$\mathbf{y}_s \sim \mathcal{P}\left(\exp\left[\sum_{h=1}^{H}(\boldsymbol{X}\boldsymbol{u}_h)\,\gamma_{s,h} + \boldsymbol{A}\boldsymbol{\delta}_s + \mathbf{Z}\boldsymbol{\xi}_s\right]\right),$$

where $\boldsymbol{\xi}_s$ is a random effect with 22 levels, modeling the dependency of abundances $\mathbf{y}_s$ within concessions. Estimating the variance of these random effects allows us to identify species for which abundances are more variable between concessions.

The optimal number of components is determined by a five-block cross-validation procedure, conserving the optimal tuning parameters from the first analysis ($\lambda = 0.15$ and $\ell = 4$). The cross-validation errors obtained with mixed-SCGLR are shown in Figure 9.6 alongside those obtained using SCGLR.

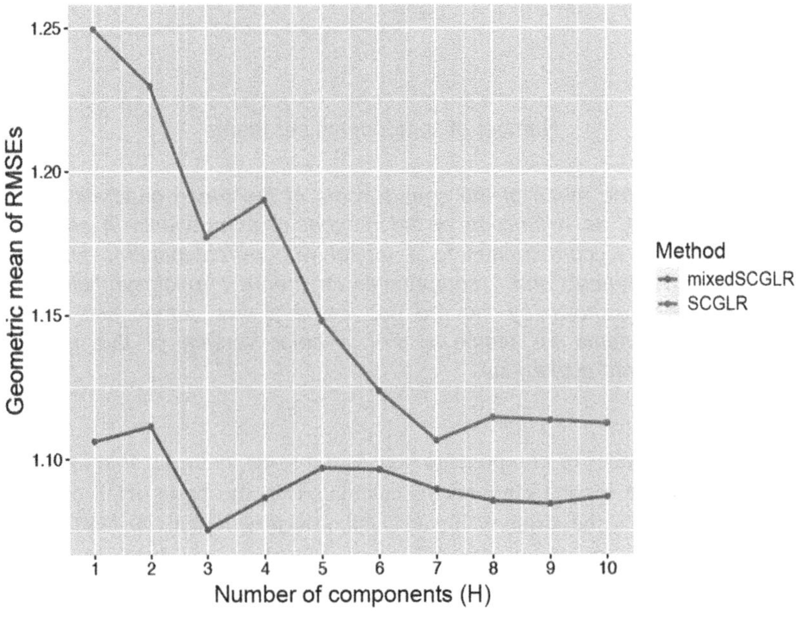

Figure 9.6. *Geometric mean of the RMSE as a function of the number of components, with $\lambda = 0.15$ and $\ell = 4$. For mixed-SCGLR, the optimal value is $H = 3$. For a color version of this figure, see www.iste.co.uk/peyrard/ecology.zip*

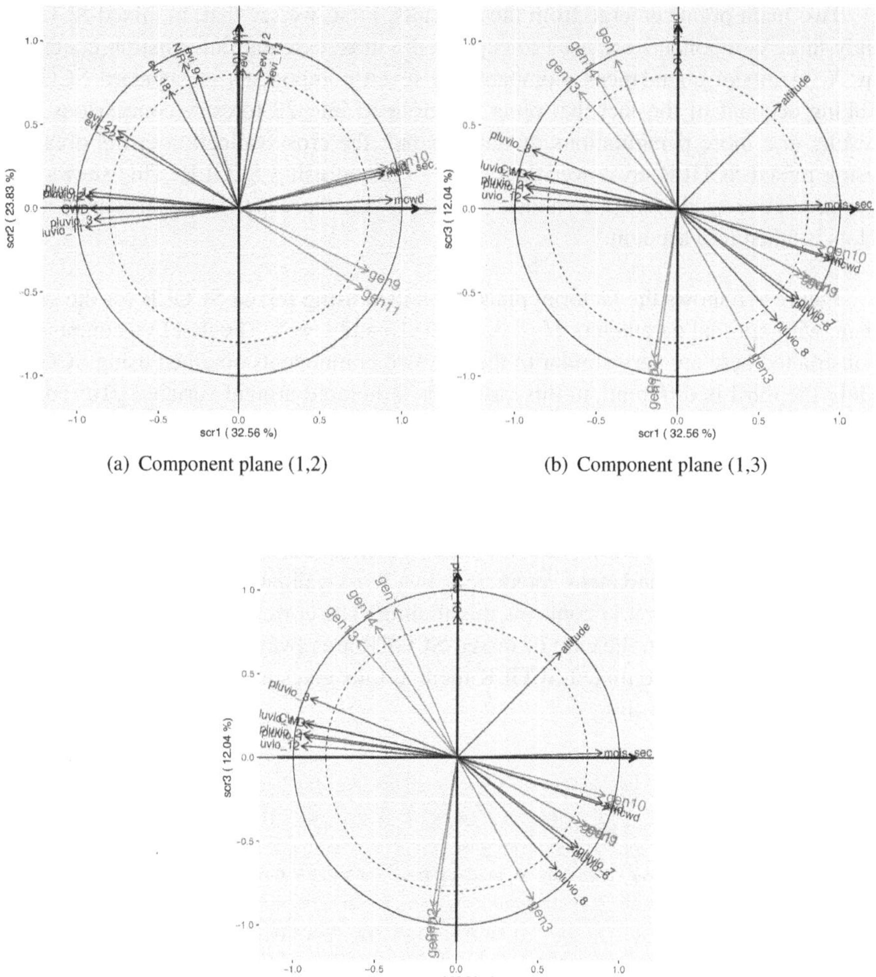

(a) Component plane (1,2)

(b) Component plane (1,3)

(c) Component plane (2,3)

Figure 9.7. *Factorial planes (1,2), (1,3) and (2,3) obtained using mixed-SCGLR on the genus dataset, with optimal parameters* $\boldsymbol{H = 3}$, $\boldsymbol{\lambda = 0.15}$ *and* $\boldsymbol{\ell = 4}$. *Variables and linear predictors are only shown for species which are sufficiently well represented (correlations above* 0.8*). For a color version of this figure, see www.iste.co.uk/peyrard/ecology.zip*

Two main points emerge from these results. First, we see that, in mixed-SCGLR, only three components are used to capture the most essential information contained in $\boldsymbol{X}$ to model abundances, compared to seven components in standard SCGLR. Taking account of the fact that plots are grouped into 22 forestry concessions thus results in a more parsimonious model. Second, the cross-validation errors obtained using mixed-SCGLR are lower than those obtained using SCGLR. This shows that the predictive quality of the model is better when the dependence structure of the plots is taken into account.

Figure 9.7 shows the factorial planes obtained using mixed-SCGLR for the *genus* data, with optimal parameters $H = 3$, $\lambda = 0.15$ and $\ell = 4$. The first two components constructed here are very similar to the first two components obtained using SCGLR. Only the third is different: in this case, it is structured around variable `pluvio_10`, whereas in SCGLR, it was structured around `altitude`.

As we see from Figure 9.6, taking account of the dependency relationship between measures taken from the same forestry concession improves the predictive quality of the model. A comparison of the Spearman correlations between observed abundances and abundances predicted by cross-validation using SCGLR and mixed-SCGLR (Table 9.1) confirms this finding. The correlations between predicted and observed values in the case of mixed-SCGLR are always greater than or equal to those for SCGLR. The mixed-SCGLR method thus gives us a finer prediction of the abundance of each species.

Species No.	1	2	3	4	5	6	7	8	9	10	11	12	13	14	15
SCGLR	0.65	0.64	0.60	0.49	0.39	0.44	0.61	0.63	0.85	0.63	0.62	0.58	0.52	0.73	0.51
Mixed-SCGLR	0.69	0.69	0.61	0.52	0.44	0.46	0.68	0.65	0.87	0.63	0.69	0.60	0.56	0.75	0.56

Table 9.1. *Spearman correlations between observed and predicted abundances obtained by SCGLR and mixed-SCGLR*

Finally, note that both SCGLR and mixed-SCGLR can be used to map predicted abundances. Figure 9.8 shows the abundance values predicted by SCGLR and mixed-SCGLR for the species with the highest inter-concession variability (species No. 12). The predictions obtained using the two methods are visually similar, but the quality is slightly higher using mixed-SCGLR (Spearman correlation of 0.6 compared to 0.58 for SCGLR).

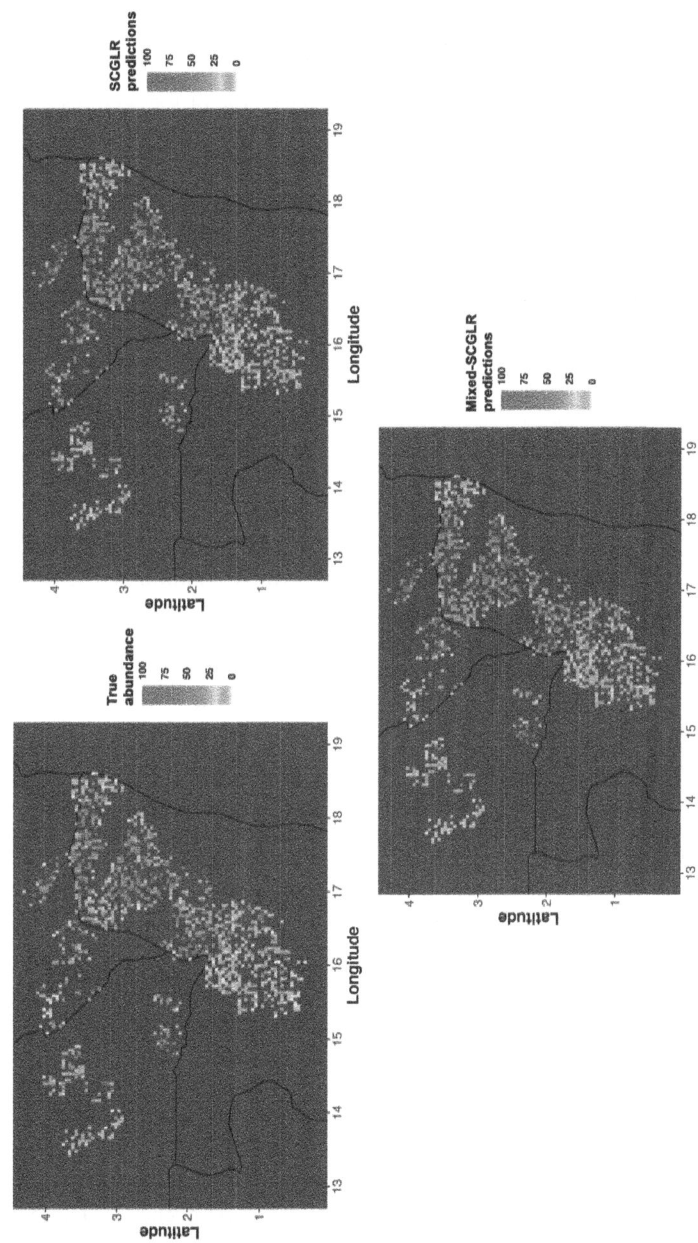

Figure 9.8. *Observed and predicted abundance maps for the species with the highest inter-concession variation in abundance (species No. 12). For a color version of this figure, see www.iste.co.uk/peyrard/ecology.zip*

9.4. Discussion

The SCGLR method was designed with three aims in mind. The first was to extract the best latent dimensions for modeling a set of responses from a set of explanatory variables. These dimensions take the form of stable and interpretable components. The second objective was to obtain a linear predictor for each response, using these components, which is not over-fitted, particularly in cases involving large numbers of explanatory variables and where some of these variables are redundant. The third and final aim was to find a way of simultaneously modeling responses with different distribution types. These aims were achieved by maximizing a criterion combining two sub-criteria: the structural relevance of the explanatory components and the goodness of fit of the model of responses based on these components. The combination of the two sub-criteria benefits from added flexibility via the use of two hyper-parameters, λ and ℓ. Parameter λ is such that, at the optimum, $\lambda/(1-\lambda)$ is the elasticity of structural relevance with respect to goodness of fit. Parameter ℓ is used to direct components toward more or less fine bundles of correlated variables. In the case of a single response variable, and when λ is equal to 0, the SCGLR method is simply a generalized linear regression. For $\ell = 1$ and when $\lambda = 1$, SCGLR corresponds to a principle component analysis. Altering the values of λ and ℓ offers the means of exploring a continuum of methods for extracting explanatory dimensions; the best model in each case is that which minimizes the mean cross-validation error. Calibrating λ and ℓ is a costly and time-consuming process. This limitation is significant, and the development of a more efficient calibration approach is an important goal for future work. Two extensions to the method were presented in sections 9.2.2 and 9.2.3. The first, THEME-SCGLR, involves thematic partitioning of explanatory variables; the resulting components, defined by theme, are conceptually clear and easy to interpret. The second variation, mixed-SCGLR, allows the use of random effects in the response model, enabling the treatment of grouped data. Further extensions may be developed in future. One option would be to consider response distributions, which are less "classic" but often encountered in real-world situations, such as zero-inflated models. Another option would be to include an elastic net type penalty in the criterion in order to obtain both parsimonious components and parsimonious linear predictors. Yet another option would be to combine SCGLR with classification algorithms in order to group responses that depend on shared explanatory components. Mixed-SCGLR offers the means of taking account of certain dependency structures between statistical units, but the range of possible dependencies goes well beyond those covered here, notably including more complex relations connected with space and time. Finally, it is important to note that SCGLR presumes the independence of responses conditional on the explanatory variables. To create a more realistic multivariate model, this approach could be extended to include a more sophisticated dependency structure for responses.

9.5. References

Aïtchison, J. and Ho, C. (1989). The multivariate Poisson-log normal distribution. *Biometrika*, 76, 643–653.

Bickel, P., Li, B., Tsybakov, A., van de Geer, S., Yu, B., Valdés, T., Rivero, C., Fan, J., van der Vaart, A. (2006). Regularization in statistics. *Test*, 15, 271–344.

Bolker, B., Brooks, M., Clark, C., Geange, S., Poulsen, J., Stevens, M., White, J.-S. (2009). Generalized linear mixed models: A practical guide for ecology and evolution. *Trends in Ecology & Evolution*, 24(3), 127–135.

Bry, X. and Verron, T. (2015). Theme: Thematic model exploration through multiple co-structure maximization. *Journal of Chemometrics*, 29, 637–347.

Bry, X., Trottier, C., Verron, T., Mortier, F. (2013). Supervised component generalized linear regression using a PLS-extension of the Fisher scoring algorithm. *Journal of Multivariate Analysis*, 119, 47–60.

Bry, X., Trottier, C., Mortier, F., Cornu, G. (2020). Component-based regularisation of a multivariate GLM with a thematic partitioning of the explanatory variables. *Statistical Modelling*, 20(1), 96–119.

Chauvet, J. (2018). Introducing complex dependency structures into supervised components-based models. PhD Thesis, Université de Montpellier, IMAG.

Chauvet, J., Trottier, C., Bry, X. (2019). Component-based regularization of multivariate generalized linear mixed models. *Journal of Computational and Graphical Statistics*, 28(4), 909–920.

Chib, S. and Winkelmann, R. (2001). Markov chain Monte Carlo analysis of correlated count data. *Journal of Business & Economic Statistics*, 19(4), 428–435.

Elith, J. and Leathwick, J. (2009). Species distribution models: Ecological explanation and prediction across space and time. *Annual Review of Ecology, Evolution, and Systematics*, 40, 677–699.

Fan, J. and Li, R. (2001). Variable selection via nonconcave penalized likelihood and its Oracle properties. *Journal of the American Statistical Association*, 96(456), 1348–1360.

Fan, J. and Li, R. (2006). Statistical challenges with high dimensionality: Feature selection in knowledge discovery. In *Proceedings of the International Congress of Mathematicians*, Sanz-Sole, M., Soria, J., Varona, J.L., Verdera, J. (eds), 3, 595–622.

Fan, J. and Lv, J. (2010). A selective overview of variable selection in high dimensional feature space. *Statistica Sinica*, 20, 101–148.

Guisan, A. and Zimmermann, N.E. (2000). Predictive habitat distribution models in ecology. *Ecological Modelling*, 135(2–3), 147–186.

Hastie, T., Tibshirani, R., Friedman, J. (2001). *The Elements of Statistical Learning*. Springer, New York.

Henderson, C.R. (1975). Best linear unbiased estimation and prediction under a selection model. *Biometrics*, 31(2), 423–447.

Hoerl, A.E. and Kennard, R.W. (1970). Ridge regression: Biased estimation for nonorthogonal problems. *Technometrics*, 12, 55–67.

Jolliffe, I. (1982). A note on the use of principal components in regression. *Applied Statistics*, 31, 300–303.

Kendall, M. (1957). *A Course in Multivariate Analysis*. Griffin, London.

Lange, K. (2012). *Numerical Analysis for Statisticians*, 2nd edition. Springer, New York.

Marx, B.D. (1996). Iteratively reweighted partial least squares estimation for generalized linear regression. *Technometrics*, 38, 374–381.

McCullagh, P. and Nelder, J. (1989). *Generalized Linear Models*, 2nd edition. Chapman & Hall/CRC Press, London.

Nelder, J.A. and Wedderburn, R.W.M. (1972). Generalized linear models. *Journal of the Royal Statistical Society, Series A*, 135, 370–384.

Pollock, L.E.A. (2014), Understanding co-occurrence by modelling species simultaneously with a joint species distribution model (JSDM). *Methods in Ecology and Evolution*, 5, 397–406.

Schall, R. (1991). Estimation in generalized linear models with random effects. *Biometrika*, 78(4), 719–727.

Tibshirani, R. (1996). Regression shrinkage and selection via the LASSO. *Journal of the Royal Statistical Society, Series B (Methodological)*, 58(1), 267–288.

Warton, D., Blanchet, F., O'Hara, R., Ovaskainen, O., Taskinen, S., Walker, S., Hui, F. (2015). So many variables: Joint modeling in community ecology. *Trends in Ecology and Evolution*, 30, 766–779.

Wilkinson, D., Golding, N., Guillera-Arroita, G., Tingley, R., McCarthy, M. (2019). A comparison of joint species distribution models for presence–absence data. *Trends in Ecology and Evolution*, 10, 198–211.

Wold, H. (1966). Estimation of principal components and related models by iterative least squares. *Multivariate Analysis*, 1, 391–420.

Zou, H. and Hastie, T.J. (2005). Regularization and variable selection via the elastic net. *Journal of the Royal Statistical Society, Series B*, 67, 301–320.

10

Structural Equation Models for the Study of Ecosystems and Socio-Ecosystems

Fabien LAROCHE[1,2], Jérémy FROIDEVAUX[1], Laurent LARRIEU[1,3] and Michel GOULARD[1]

[1] *UMR 1201 Dynafor, University of Toulouse, INRAE INPT EI PURPAN, Castanet-Tolosan, France*

[2] *INRAE, UR EFNO, Nogent-sur-Vernisson, France*

[3] *CNPF-CRPF Occitanie, Tarbes, France*

10.1. Introduction

10.1.1. *Ecological background*

An ecosystem is a complex entity made up of a large number of interacting, heterogeneous biological and physical components. Human societies make a living by using the ecosystems in which they exist, notably via agriculture and forestry. This combination of a human society and one or more ecosystems on which the society depends is known as a socio-ecosystem.

The generalization of productivist logic in humanity's relationship with ecosystems has led to the erosion of certain components of agricultural and forest ecosystems (e.g. the level of organic matter in farmland, or the quantity of deadwood in a forest) and to a decline in biodiversity. This decline raises ethical issues, in terms of conservation, and reduces the quality and diversity of benefits that humanity can derive from these ecosystems; thus, the socio-ecosystem itself is becoming increasingly fragile.

Statistical Models for Hidden Variables in Ecology,
coordinated by Nathalie PEYRARD and Olivier GIMENEZ. © ISTE Ltd 2022.

Before we can identify changes to management and usage practices with the potential to correct this erosion effect, we need to understand the causality connections between these practices, the physical characteristics of the milieus in question and the biodiversity of these sites.

The spatial extent and complex nature of socio-ecosystems make them difficult to study under controlled conditions. The causality connections between components can only be interpreted using a comparative approach; in statistical terms, this entails the identification of quantitative relationships between certain characteristics.

10.1.2. *Methodological problem*

The data sets used to study relationships between components of socio-ecosystems often comprise a relatively small quantity of independent ecosystems (the *observations*) and a large quantity of data for each of these ecosystems (the *variables*). The addition of a new socio-ecosystem implies the acquisition of a large number of measures of different natures, whereas the addition of a new variable for socio-ecosystems that are already being monitored is generally relatively easy, particularly if the data already exist and must simply be recomposed.

The search for relationships within a large group of variables may require multiple tests to be applied to the same dataset. Without adequate control, the risk for the first species of each test, taken individually, may be inflated. However, existing control methods (Holm 1979; Benjamini and Hochberg 1995) have a detrimental effect on the power of tests, which may already be low due to a small number of observations. Separate analysis of the various relationships that are expected to exist in the model, for example using regression models, may thus result in either false positives or too few detections.

The components of interest in socio-ecosystems, such as biodiversity, exploitation practices, and so on are inherently complex in nature and can only be studied indirectly, through the lens of groups of measured variables within a dataset. The notion of plant biodiversity in a managed forest landscape, for example, is partly reflected in measures such as the number of plant species observed in a sampling process, the number of tree species, the number of bryophyte species, and so on. Causality connections between these components may be analyzed by studying the relationships present within a reduced number of groups of variables, rather than by considering a large number of variables individually. Using this grouped structure may result in improved detection of the relationships between components by reducing the number of relationships to explore. Furthermore, the level of redundancy between variables in the same group is likely to reduce error in characterizing the components of socio-ecosystems, resulting in a clearer view of the relationships between these components.

The notion of latent variables may be used to model components in a socio-ecosystem, as described above. Latent variables offer a means of drawing on the grouped structure of observed variables in a socio-ecosystem and of avoiding the methodological problems described earlier. Structural equation models (SEMs) are a practical implementation of this idea.

10.1.3. *Case study: biodiversity in a managed forest*

At a global level, forests are home to around two-thirds of terrestrial biodiversity. Most forests in Europe have long been managed, and this has major implications for their biodiversity. Following the 1992 Rio Convention, forest managers are now required to protect species diversity in harvested areas. However, species inventories and taxonomic expertise come at a cost, and cannot be carried out on a regular basis in all forests. This raises the need for indirect approaches to evaluating biodiversity on a level compatible with forest management practices, making it easier to take account of biodiversity on a day-to-day basis.

A recent study on mainland France (Larrieu *et al.* 2019) focused on quantifying the relationships between descriptive, directly measurable variables for forests and samples of local biodiversity. These relationships have the potential to highlight variables that may act as indirect indicators of local biodiversity. The study was based on a relatively large dataset (487 observations, each corresponding to a circular area of 1 hectare), comparing descriptive variables with sampling data across a range of different forest contexts, for a number of species that varied between observations. The taxonomic richness of different groups and the descriptive variables in the dataset were all considered as components in the analysis (Larrieu *et al.* 2019). Each biodiversity variable is linked to predictors, observed using a classic regression model. The methodological issues highlighted above do not arise in this case due to the quantity of available data. In this chapter, the SEM approach will be used to integrate the grouped structure of observed variables into this analysis. In order to maximize the number of species groups which can be treated simultaneously and to maximize the homogeneity of population types, we have chosen to work on a sub-set of the data from (Larrieu *et al.* 2019), consisting of 41 observations from two low-altitude, predominantly deciduous forests.

Fifteen descriptive variables are available for each observation site, comprising historical, structural and composition variables that are considered particularly relevant for forest contexts. Some of these variables are directly linked to the population and current management practices, characterizing vegetation, deadwood, very large trees, habitat trees (which contain unique structural elements, known as dendromicrohabitats, which are home to certain species) and open spaces. Other variables relate to the history and context of the forest: the temporal continuity of the wooded area, and the presence of rocky and aquatic features (for more information concerning the ecological and forest management aspects of these variables, see Larrieu and Gonin 2008).

The data also include biodiversity measures, obtained by sampling at the center of the site. These measures relate to several groups of species. Three groups are intrinsically linked to woody substrate and deadwood: polypores, saproxylic beetles (which rely on decomposing wood or other saproxylic species for at least part of their life cycle) and bryophytes (corticolous or saproxylic). Two groups depend on the presence of specific structures in trees (dendromicrohabitats like cavities, peeling bark, etc.): bats and cavity-nesting birds. Three further groups that are associated with the forest environment without clearly depending on the presence of the attributes listed above include vascular woodland plants, non-cavity nesting woodland birds and non-cavity nesting bats. The final three groups – ground beetles, non-forest specialist vascular plants and non-forest specialist birds – have a broader ecology and are not necessarily dependent on the woodland environment. One or more biodiversity measures are used for each sampled group, giving a total of 13 useable measures of species richness.

10.2. Structural equation model

As we saw at the end of section 10.1.2, our aim here is to model the components of interest in a socio-ecosystem in the form of latent variables, in the sense described in the introduction to this book. Observed variables will be considered as measures of the latent variables in the model. An SEM consists of (i) the relations between each latent variable and the observed variables used to measure them (forming groups of observed variables), and (ii) the relationships between latent variables. The set of type (i) relationships constitutes the *measurement model*, while the type (ii) relationships make up the *relational model* (see Figure 10.1). The relational model, which describes the relationships between components in a socio-ecosystem, is generally the focus of most studies.

10.2.1. *Hypotheses and general structure of an SEM*

10.2.1.1. *Data set*

Consider a data set made up of u observations of n variables. These variables are presumed to be centered and reduced, such that the empirical variance–covariance matrix S of the observed variables is a correlation matrix. The data set is modeled as the result of u independent drawings of n variables following a multivariate Gaussian distribution. The marginal distributions are presumed to be centered and reduced such that the associated variance–covariance matrix, denoted as Σ, is a correlation matrix. In other terms, the fact that the data set is centered and reduced is used to establish the actual means and variances of the observed variables *a priori*. This hypothesis is used for reasons of commodity, but is not strictly necessary (see section 10.5).

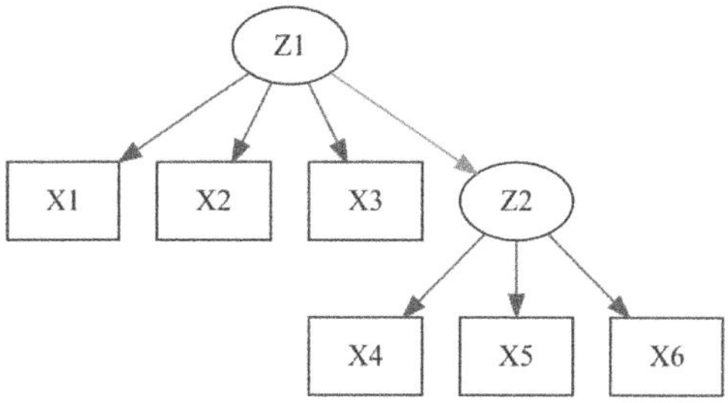

Figure 10.1. *Conventions used in representing an SEM. Latent variables are shown as circles, while observed variables are represented by squares. Relationships are indicated by arrows. Blue arrows connecting a latent variable and an observed variable indicate that this relationship belongs to the measurement model. Red arrows between latent variables indicate relationships in the relational model. In this case, the relational model includes just one relationship, whereas the measurement model contains six relationships. In the example shown here, X_1, X_2 and X_3 form a group of observed variables that measure the latent variable Z_1, and Z_1 has a causal effect on Z_2. For a color version of this figure, see www.iste.co.uk/peyrard/ecology.zip*

10.2.1.2. *Formalization of the measure model and relational model*

10.2.1.2.1. Measurement model

We shall consider that the observed variables are linked to p latent variables $Z_1, ..., Z_p$ with centered reduced Gaussian distributions. Here, we presume that each observed variable is connected to just one latent variable. If an observed variable X_k is connected to a latent variable Z_i, then:

$$X_k = \alpha_{ik} Z_i + E_k \quad [10.1]$$

where $\alpha_{ik} \in [0, 1]$, and E_k is a centered Gaussian variable of variance $1 - \alpha_{ik}^2$. Variables E_k represent measurement errors, and are taken to be pairwise independent (both within a group and between groups of observed variables associated with different latent variables). The set of equations of type [10.1] defines the measurement model of the SEM. Coefficients of type α_{ik} can be compiled into a rectangular matrix A of dimension $p \times n$ such that $A_{ik} = \alpha_{ik}$ if the observed variable X_k measures the latent variable Z_i and $A_{ik} = 0$ otherwise.

Matrix A enables relations of type [10.1] to be summarized in a compact manner:

$$\mathbf{X} = t(A)\,\mathbf{Z} + \mathbf{E} \quad [10.2]$$

where $t(M)$ is the transpose of matrix M, $\mathbf{X} = t(X_1, ..., X_n)$, $\mathbf{Z} = t(Z_1, ..., Z_p)$ and $\mathbf{E} = t(E_1, ..., E_n)$.

10.2.1.2.2. Relational model

The relational models used in SEM must be acyclic: there must be no paths within the oriented graph of relations between latent variables (red arrows according to the convention set out in 10.1) leading back to the starting point (see Pearl 2000, p. 13).

The relationships between latent variables in the relational model (the structural equations) are written as follows:

$$Z_j = \sum_{i=1}^{p} \beta_{ij} Z_i + F_j, \qquad [10.3]$$

where $\beta_{ij} \in \mathbb{R}^*$ if Z_i is presumed to have a causal effect on Z_j (i.e. there is an arrow from Z_i to Z_j in the graphical representation of the model: see Figure 10.1) and $\beta_{ij} = 0$ otherwise. F_j is a random Gaussian variable with a mean of zero and variance δ_j^2 describing the variations in Z_j, which are not explained by the relational model. Thus, in a relational model with p latent variables, we have p residual variance terms $F_1, ..., F_p$. The causal effects between latent variables reflected in coefficients of type β_{ij} may be compiled into a square matrix B of dimensions $p \times p$ such that $B_{ji} = \beta_{ij}$. As mentioned before, B must be interpretable as the matrix of links in a weighted acyclic graph. Using matrix B, relations of type [10.3] can be expressed in a compact manner:

$$\mathbf{Z} = B\mathbf{Z} + \mathbf{F}, \qquad [10.4]$$

where $\mathbf{Z} = (Z_1, ..., Z_p)$ and $\mathbf{F} = (F_1, ..., F_p)$

Terms F_1, ..., F_p are presumed, by default, to be pairwise independent. However, factors that are not captured by the observed variables, and are thus not represented in the latent variables, may generate correlations between terms. If two terms F_i and F_j are correlated in this way, a parameter γ_{ij} is used to describe their covariance. These parameters are then compiled into a covariance matrix Σ_F such that: (i) $[\Sigma_F]_{ii} = \delta_i^2$; (ii) $[\Sigma_F]_{ij} = 0$ if F_i and F_j are not presumed to be correlated; (iii) $[\Sigma_F]_{ij} = \gamma_{ij} \in [-\delta_i\delta_j, \delta_i\delta_j]$ if they are believed to be correlated.

The hypothesis that the latent variables are standardized and equation [10.3] induce certain constraints for parameters β_{ij}, δ_j and γ_{ij}:

$$\mathbb{V}(Z_j) = 1 = \left[\sum_{i=1}^{p} \beta_{ij}^2 \mathbb{V}(Z_i) + \delta_j^2\right] + 2\left[\sum_{i_1=1}^{p-1} \sum_{i_2=i_1+1}^{p} \beta_{i_1,j}\beta_{i_2,j}\mathbb{C}ov\left(Z_{i_1}, Z_{i_2}\right)\right] \qquad [10.5]$$

A latent variable Z_i that is not causally affected by any other latent variables in the model gives us $Z_i = F_i$. Variables of this type are known as *exogenous variables*. The type of decomposition shown in equation [10.5] can be applied recursively over the variance and covariance terms of the right member in order to obtain the exogenous variables (see Table 16.2 of Sokal and Rohlf 1995). The variance and covariance terms of exogenous variables are known in advance: the variances correspond directly to the residual variances δ_j^2 and the variable reduction constraint implies $\delta_j^2 = 1$; the covariances are either zero, or correspond to parameters of the type γ_{ij}. This finally gives us a constraint on the parameters β_{ij}, δ_j and γ_{ij} used in chains of causal relations ending with Z_j. By repeating this operation for all latent variables of the model, we obtain p constraints on the parameters.

10.2.2. *Likelihood and estimation in an SEM*

All of the tools used with a classic linear model can be applied in order to estimate and test SEMs. In this chapter, we shall focus on estimation and testing approaches based on the likelihood of the parameters of the SEM, and on likelihood relations between SEMs. Other approaches (least squares, Bayesian estimation, etc.; see Lei and Wu 2015) exist, but the statistical framework is generally less sophisticated.

10.2.2.1. *Likelihood of parameters* (A, B, Σ_F) *in an SEM*

The hypothesis that the observed variables obey a centered multivariate Gaussian distribution implies that the log-likelihood of the parameters (A, B, Σ_F) of an SEM corresponds to the log-likelihood of the correlation matrix Σ, which they predict for the observed variables:

$$LL\left(\Sigma; [x]\right) = C - \tfrac{u}{2}\left[\log\left(|\Sigma|\right) + \tau\left(\Sigma^{-1}S\right)\right], \qquad [10.6]$$

where $[x]$ denotes the table of observations of dimension $u \times n$, $C = -\frac{un}{2}\log\left(2\pi\right)$ is not dependent on the values of the data or on the parameters of the SEM, $|\Sigma|$ denotes the determinant of Σ and $\tau\left(\Sigma^{-1}S\right)$ denotes the trace of $\Sigma^{-1}S$.

Equation [10.6] is a classic result in the linear model (Rao 1973; Muirhead 1982), which is at the heart of SEM theory (see Appendix 4a of Bollen 1989). This implies that the likelihood of the basal parameters of an SEM is only dependent on the data via the adequation between the induced correlation matrix Σ and the empirical correlation matrix S.

10.2.2.2. *Maximum likelihood estimator in an SEM*

Working on the hypothesis that B is the matrix of an oriented weighted acyclic graph, the correlation matrix Σ can be expressed as a function of the basal parameters of the SEM, obtained by combining [10.2] and [10.4]:

$$\Sigma = \Sigma\left(A, B, \Sigma_F\right) = t\left(A\right)\left[\sum_{s=1}^{p} t\left(B\right)^s \Sigma_F B^s\right] A + \Delta_{1-\alpha^2}, \qquad [10.7]$$

where $\Delta_{1-\alpha^2}$ is a diagonal matrix of dimension $n \times n$ of which the kth diagonal coefficient is worth $1 - \alpha_{ik}^2$, where α_{ik} is the coefficient linking the observed variable X_k to the latent variable Z_i, which it measures.

In practice, finding the maximum likelihood estimator of an SEM comes down to maximizing the following criterion:

$$\lambda(A, B, \Sigma_F) = -\left[\log(|\Sigma(A, B, \Sigma_F)|) + \tau\left(\Sigma(A, B, \Sigma_F)^{-1} S\right)\right]$$

10.2.2.2.1. Parameter constraints

Maximization is carried out under a set of constraints of various origins: (i) the structure of the SEM means that certain coefficients of A, B and Σ_F must be zero; (ii) the reduced character of the latent and observed variables imposes constraints of the type [10.5], and the coefficients of type α must have an absolute value of less than 1; (iii) the coefficients of type α must be positive; (iv) matrix Σ_F must be symmetrical and positive definite.

In practice, it is hard to guarantee that the matrix Σ_F will remain positive definite throughout the optimization process, particularly if the model contains free correlation parameters of type γ. Our approach is to optimize the likelihood without the positive definite constraint on Σ_F, then to project the estimator of Σ_F obtained into this way onto the space of positive definite matrices, before re-estimating the other parameters conditional on this projection.

10.2.2.2.2. Identifiability problem

Certain structures in the measurement model and relational model of SEMs can result in situations in which any matrix Σ, which can be attained by equation [10.7], can also be attained using an infinite number of combinations of distinct basal parameters (A, B, Σ_F). In this case, the maximum likelihood estimator is not defined in a unique manner, and numerical optimization methods will not converge: this is known as an identifiability problem. For example, in an SEM with a single latent variable measured by a group of two observed variables, there are an infinite number of combinations of parameters α with the same product, which will result in maximum likelihood.

This type of problem in measurement models can be avoided through the use of two additional constraints: (i) if a latent variable Z_i only takes a single observed variable X_k as the measure, then α_{ik} is arbitrarily fixed at 1; (ii) if a latent variable Z_i can take two observed variables X_k and X_l as measures, and is not itself involved in a relationship in the relational model (i.e. line i and column i in B are zero), then we take α_{ik} or α_{il} fixed at 1. It is harder to avoid identifiability problems connected to the structure of the relational model. One necessary condition is that the number of arrows in the overall SEM must not exceed the number of empirical correlation coefficients, equal to $\frac{n(n-1)}{2}$ (see Grace 2006, pp. 116-121, for more details); however, this condition is not sufficient.

10.2.3. *Fit quality and nested SEM tests*

Under the hypothesis of identifiability, the classic theory of the linear model indicates that the maximum likelihood estimator is almost certain to exist. We now wish to examine the fit of the SEM estimated using maximum likelihood. Let $\hat{\Sigma}$ be the correlation matrix of variables associated with this SEM.

10.2.3.1. *Goodness-of-fit test*

The quality of an estimated SEM can be assessed by means of a goodness-of-fit (GoF) test, designed to choose between the null hypothesis $H_0 : \Sigma = \hat{\Sigma}$ and the alternative hypothesis $H_1 : \Sigma \neq \hat{\Sigma}$. Rejection of H_0 implies that the estimated SEM is not sufficient to describe the correlations between observed variables (i.e. arrows need to be added to the graph). This test offers a means of explicitly checking the type I error rate, which is the probability of falsely concluding that the estimated SEM is not sufficient to describe the data (i.e. concluding that H_1, while in reality, H_0 is true). However, it does not guard against type II errors, the probability of falsely concluding that the SEM in question is adequate (i.e. concluding that H_0 when in fact H_1 is true). This latter risk may be significant if not controlled, and acceptance of the null hypothesis H_0 is only weak proof of the validity of the estimated SEM.

To implement this test, we can consider the statistics of the difference in log-likelihood: $\Delta_{LL} = LL\left(\hat{\Sigma}; [x]\right) - LL\left(S; [x]\right)$. Δ_{LL} is calculated using equation [10.6].

10.2.3.1.1. Asymptotic test

Under asymptotic hypotheses (i.e. for very high values of u) and in the identifiable case, $-2\Delta_{LL}$ is distributed according to a χ^2 distribution with r degrees of freedom under H_0 (Lejeune 2010, p. 225), where r is the number of arrows to add to the estimated SEM in order to obtain an SEM with the capacity to attain any correlation matrix Σ, while remaining identifiable. In practice, r may be evaluated by determining the number f of arrows (across both the measurement model and the relational model) in the graphic representation of the estimated SEM, taking $r = \frac{n(n-1)}{2} - f$. H_0 is rejected if the quantile q corresponding to the statistic Δ_{LL} observed in a $\chi^2(r)$ distribution is greater than $1 - a$, where a is the level desired for the test (generally, $a = 0.05$). The p-value of the test is $1 - q$.

10.2.3.1.2. Re-sampling test

The p-value of the test based on a χ^2 distribution of $-2\Delta_{LL}$ under H_0 is one of the most widespread measures used to evaluate the quality of SEMs (despite the weakness of the validation). However, the χ^2 distribution of $T = -2\Delta_{LL}$ under H_0 is a result which requires a high number of observations u, and this is rarely the case in our context of study. Re-sampling (bootstrap) methods are another recommended option for approximation the distribution of $T = -2\Delta_{LL}$ under H_0.

The classic non-parametric bootstrap approach consists of a random drawing with replacement of u observations from the group of initial observations, and recalculating T for the sample obtained in this way. By repeating the operation a high number B of times, we obtain a series of values of T: T_1^*, T_2^*,..., T_B^* conditional on the empirical distribution of observations. However, the distribution of T^* generated in this way is very similar to an $H_0 : \Sigma = S$ hypothesis, rather than the desired $H_0 : \Sigma = \hat{\Sigma}$ hypothesis. It is thus likely to be closer to the observations than the result of the $H_0 : \Sigma = \hat{\Sigma}$ hypothesis, and the type II error rate may be particularly high (Bollen and Stine 1992).

To correct this issue, Bollen and Stine (1992) propose the application of a non-parametric re-sampling approach to transformed observations. In this case, we prefer a more direct approach based on parametric re-sampling: sets of re-sampled observations are generated via simulations of a centered reduced multivariate Gaussian distribution with a correlation matrix $\hat{\Sigma}$. H_0 is rejected if the quantile $\hat{q}$ corresponding to T in the distribution of the B values of T^* generated by the simulations is significantly greater than $1 - a$, where a is the level desired for the test.

10.2.3.2. *Comparing two nested SEMs using likelihood ratios and re-sampling*

The likelihood ratio test presented above evaluates whether or not the relationships included in the SEM are sufficient to give an adequate description of the data. Similarly, we may wish to test whether all of the relationships included in the SEM are useful. To do this, the SEM in question must be compared with a sub-model in which a certain number of relations have been eliminated (i.e. set at 0). Two nested models can be compared using a likelihood ratio test based on the test statistic $\Delta_{LL}^{1,2} = LL\left(\hat{\Sigma}_1; S\right) - LL\left(\hat{\Sigma}_2; S\right)$, where $\hat{\Sigma}_1$ denotes the correlation matrix induced by the maximum likelihood estimator for the most constrained SEM, and $\hat{\Sigma}_2$ is that obtained with the least-constrained SEM.

Once again, we recommend using a parametric re-sampling approach, consisting of (i) a high number of simulations of sets of observations, distributed according to a multivariate Gaussian distribution with a correlation matrix equal to $\hat{\Sigma}_1$, calculating the statistic $\Delta_{LL}^{1,2*}$ each time; (ii) rejecting H_0 if the quantile of $-2\Delta_{LL}^{1,2}$ in the series of $-2\Delta_{LL}^{1,2*}$ resulting from the simulations is significantly higher than $1 - a$ where a is the level desired for the test.

The fact that the type II error rate cannot be controlled is less problematic in this case than for the global SEM test presented in section 10.2.3.1.2. In the global SEM adequation test, the validation of the estimated SEM was associated with the acceptance of H_0, a decision for which the potential error (second species risk) was not controlled. Here, the validation of the estimated SEM is clearly associated with the rejection of H_0, in which case the potential error (type I error rate) is well controlled.

10.3. Case study: biodiversity in managed forests

10.3.1. *Preliminary steps*

Variables from the data set described in section 10.1.3 are transformed to obtain centered-reduced Gaussian variables by means of power transformations (Box and Cox 1964). Seven variables, for which the transformation is fruitless, are eliminated: this leaves us with a data set of 41 observations and 21 variables (see Table 10.1). A measurement model is constructed based on latent variables relating to biodiversity and describing populations, corresponding to the broad categories defined in section 10.1.3. Some of these categories are subdivided so that the observed variables within a group are all positively inter-correlated (see Table 10.2). A relational model based on *a priori* knowledge is then introduced. For example, the presence of very large trees (*TGB*) affects the availability of dendromicrohabitats (*DMH*), as features such as cavities are mostly found in trees of large diameter. Finally, we obtain the SEM shown in Figure 10.2.

The next stage is to test the fit of the SEM proposed in Figure 10.2 with the data, following the three main steps illustrated in Figure 10.3.

10.3.2. *Evaluating the measurement model alone*

This test aims to determine whether the structure of the measurement model is compatible with the data, independent of the relational model (see Figure 10.3, test 1). This is done using *confirmatory factor analysis* (see Bollen 1989, pp. 226–318). Let us consider an SEM with the targeted measurement model (blue arrows in Figure 10.2), but in which all possible correlation structures between latent variables are permitted. If this model is rejected by the GoF test (section 10.2.3.1.2), then the structure of the measurement model is inadequate, and the definition of the latent variables and their connections to observed variables needs to be revised. This parametric re-sampling test is then applied in our example. The empirical quantile of the statistic $-2\Delta_{LL}$ observed in the re-sampled distribution ($B = 500$ repetitions) is worth $\hat{q} = 0.458$, implying that the actual quantile of $-2\Delta_{LL}$ is less than 0.95 under H_0. Thus, the confirmatory factor analysis SEM will not be rejected by a test with a level of 0.05, and the proposed measurement model will be accepted as the result of the first step in Figure 10.3.

Observed variable	Definition
rs.carab	Number of ground beetle species
rs.colsaproflor	Number of saproxylic beetle species, flower-dwelling at adult stage
rs.bryo	Number of corticolous or saproxylic bryophyte species
rs.chirocav	Number of cavicolous forest bat species
rs.oisocav	Number of cavicolous forest bird species
rs.myco.d	Number of saproxylic polypore species
rs.florafor	Number of vascular forest plant species
rs.oisoForNCav	Number of non cavicolous forest bird species
rs.colsaproNFlor	Number of non flower-dwelling saproxylic beetles
rs.floraNFor	Number of non-forest vascular plants
rs.myco.ND	Number of saproxylic fungus species other than polypores
nb.ess	Number of tree species on site
nb.chand	Number of standing deadwood (diam. $\geq$ 40cm) on site
nb.bms	Number of pieces of deadwood on ground (diam. $\geq$ 40cm) in site
nb.tgb	Number of very large living trees (diam. $\geq$ 70cm) on site
nb.cav	Number of trees with cavities on site
nb.unbark	Number of trees with exposed (barkless) wood on site
nb.bmh	Number of trees with deadwood in the crown on site
open	Proportion of ground covered with flowering plants (mostly helophiles)
bufPres	Proportion of forest coverage in a radius of one kilometer around the site at time of collection
bufPast	Proportion of forest coverage in a radius of one kilometer around the site in 1850

Table 10.1. *Variables used in the case study*

Latent variable	Definition
BIOSAPRO	Saproxylic biodiversity, dependent on wood substrate or deadwood
BIOBRYO	Biodiversity of bryophytes
BIODMH	Biodiversity dependent on dendromicrohabitats
BIOFOR	Other forest biodiversity
BIOFLORANFOR	Biodiversity of flora that does not grow under a full forest canopy
BIOCARAB	Biodiversity of ground beetles, not forest specialists
DIVESS	Biodiversity of tree species
OPEN	Openness of the stand
TGB	Presence of very large trees in the stand
BUF	Stand belonging to a spatial and temporal continuity of forest cover
BM	Availability of deadwood in the stand
DMH	Availability of dendromicrohabitats in the stand

Table 10.2. *Latent variables based on prior knowledge*

10.3.3. *Evaluating the relational model*

Reasoning is now conditional on the structure of the measurement model identified by confirmatory analysis.

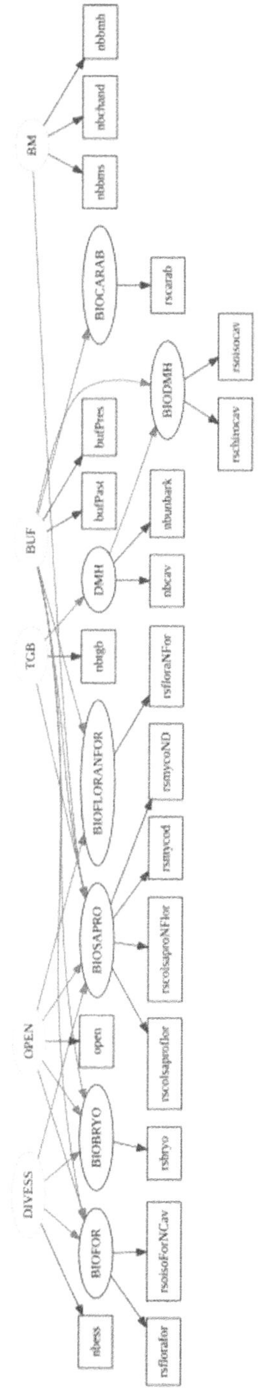

Figure 10.2. *SEM defined a priori for our case study. The exogenous latent variables (green border) can take any correlation structure (i.e. a coefficient γ may exist between each pair). For a color version of this figure, see www.iste.co.uk/peyrard/ecology.zip*

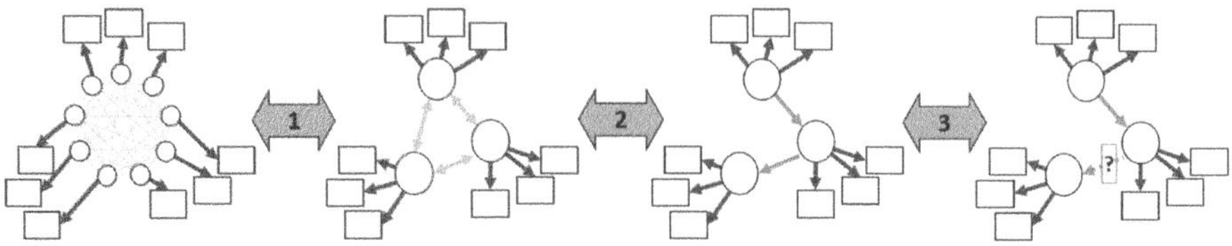

Figure 10.3. *Main steps in SEM analysis. The diagrams follow the SEM presentation approach illustrated in Figure 10.1. The green double arrows correspond to free correlations, which are permitted to exist in the relational model (γ terms). Each orange double arrow indicates a type of test, corresponding to a main step in the analytical process. Test 1: evaluation of the measurement model alone using confirmatory factor analysis (CFA). Test 2: evaluation of the relational model by comparison with the selected CFA model. Test 3: significance of a relationship within the selected relational model, using a nested model test with or without the target relationship. For a color version of this figure, see www.iste.co.uk/peyrard/ecology.zip*

10.3.3.1. *Convergence of estimation*

Figure 10.4 shows the estimators obtained for the SEM in Figure 10.2. The figure exhibits divergence phenomena, associated with the fact that highly correlated latent variables are used simultaneously to predict the same target latent variable. For example, the effect of the availability of deadwood in a stand (*BM*) on saproxylic biodiversity (*BIOSAPRO*) is close to 2, while, at the same time, the effect of belonging to a spatially and temporally continuous forest environment (*BUF*) is less than -1 (see Figure 10.4, section A). The presence of effects with an absolute value greater than 1 in a standardized relational model is not impossible, but is highly unlikely (such a result would suggest that the predictor in question explains over 100% of the variance of the target variable on its own). In this example, the extreme effects appear to result from the fact that the latent variables *BM* and *BUF* have a very high estimated correlation (see Figure 10.4, section B; correlation value 0.87).

Associations of highly correlated latent predictors are then eliminated from the model, based on section B in Figure 10.4. This results in a new SEM, shown in Figure 10.5, in which none of the parameter estimators β have an absolute value in excess of 1.

10.3.3.2. *Comparison with confirmatory factor analysis*

The SEM in Figure 10.5 is compared with the CFA SEM obtained in section 10.3.2 using a nested model comparison test based on parametric resampling (see Figure 10.3, test 2). This test evaluates whether the chosen relational model is sufficient to express the correlation structure between latent variables. An empirical quantile of $\hat{q} = 0.98$ is obtained for $B = 500$ simulated samples, which suggests that our relational model is not sufficient.

Possible sources of deviation between the SEM and the CFA are analyzed by visualizing the divergences between the correlation matrices of the latent variables (Σ_Z; Figure 10.6) predicted by these two models. Using this visualization, it is possible to identify relations that could be added to the relational model in order to reduce the degradation of fit with respect to the CFA model.

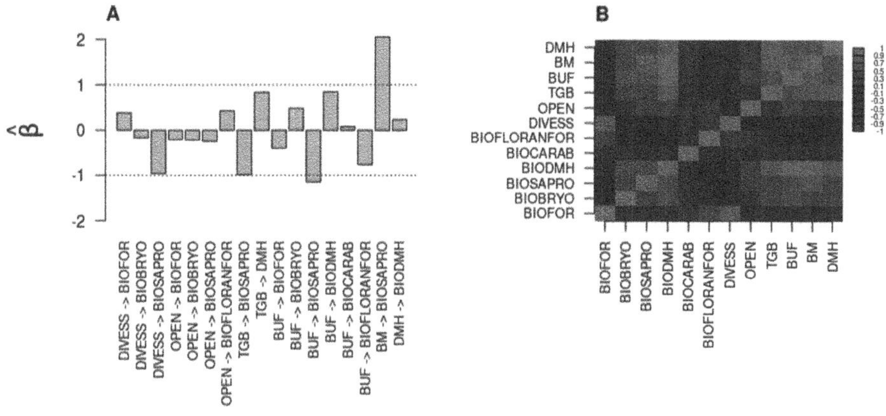

Figure 10.4. *Estimators of the parameters of the SEM in Figure 10.2. Section A: β type parameters of the relational model. Section B: Estimated correlations between the latent variables in the model (corresponding to γ type parameters for exogenous variable pairs). For a color version of this figure, see www.iste.co.uk/peyrard/ecology.zip*

Candidate relationships are selected based on theoretical hypotheses concerning the functioning of the forest ecosystem. For example, adding a relation from *BIOSAPRO* to *BIODMH* in the model may reinforce the correlation between saproxylic biodiversity (*BIOSAPRO*) and biodiversity, which is dependent on dendromicrohabitats (*BIODMH*), a need which was clearly identified in Figure 10.6(B). From an ecological perspective, it is not unreasonable to suppose that stands with a greater diversity of saproxylic species are also attractive, in terms of food resources, to a wide range of organisms that rely on the availability of dendromicrohabitats. This relationship is thus added to the model. Similarly, a relationship is added to reflect the effect of the availability of deadwood at a site (*BM*) on the biodiversity of bryophyte (*BIOBRYO*) species; this corresponds to the saproxylic character of many of these species. The same approach is taken for each of the relationships suggested in Figure 10.6. At the same time, relations with a very low estimator β (< 0.05) are eliminated from the relational model in Figure 10.5 to compensate for the possible loss of parsimony.

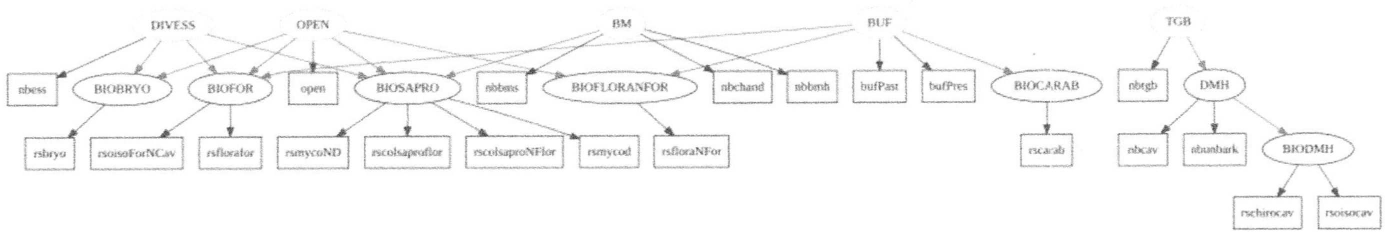

Figure 10.5. *SEM resulting from correction of the model shown in Figure 10.2, after noting the divergence of certain estimators. For a color version of this figure, see www.iste.co.uk/peyrard/ecology.zip*

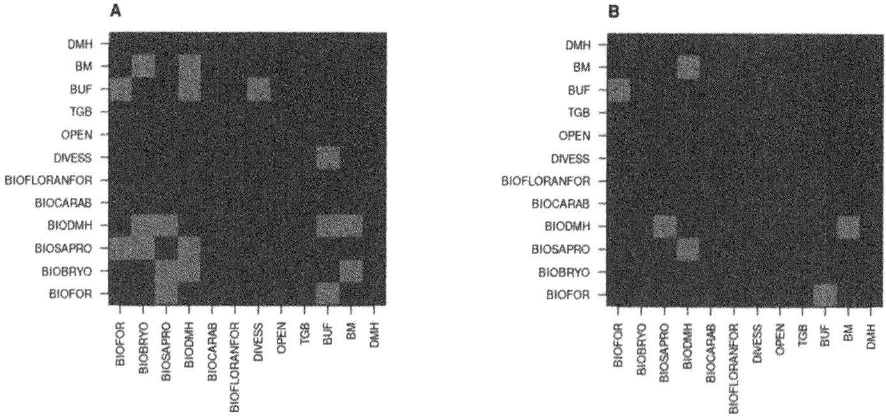

Figure 10.6. *Divergences between the correlation matrices of the latent variables (Σ_Z) of the SEM in Figure 10.5 and the confirmatory analysis model. Divergences are quantified coefficient by coefficient, testing whether the correlation predicted by CFA deviates from the SEM prediction (approximate correlation comparison test; see Lejeune 2010, p. 241). Pairs of latent variables for which the correlation predicted by CFA is significantly higher or lower than that predicted by the SEM are shown in red and blue, respectively, for a threshold of 0.05 (section A), or after conservative correction using the Bonferronni method to take account of multiple tests (section B). For a color version of this figure, see www.iste.co.uk/peyrard/ecology.zip*

The resulting SEM is shown in Figure 10.7. This new model is not rejected by the parametric resampling process MES ($\hat{q} = 0.81$, which is significantly lower than previous value of 0.95 for $B = 500$ simulated samples), and the anomalous estimation values have been removed (see Figure 10.8). Furthermore, this model now performs as well as the CFA SEM using the nested model comparison test based on parametric resampling: the empirical quantile of Δ_{LL}^{12} over $B = 500$ simulated samples is $\hat{q} = 0.89$, significantly lower than 0.95. This final SEM will thus be taken as the result of the second step shown in Figure 10.3. Note that an asymptotic fit test would have given a different conclusion in this case, and the model would have been rejected with a value of $p = 4 \times 10^{-6}$.

10.3.4. *Significance of parameters in the relational model*

The components of the relational model of the SEM selected at the end of the previous section (see Figure 10.7) are shown in Figure 10.8.

The significance of one or more parameters of type β is tested using the approach described in section 10.2.3.2, that is, a test of the likelihood ratio by resampling between nested models.

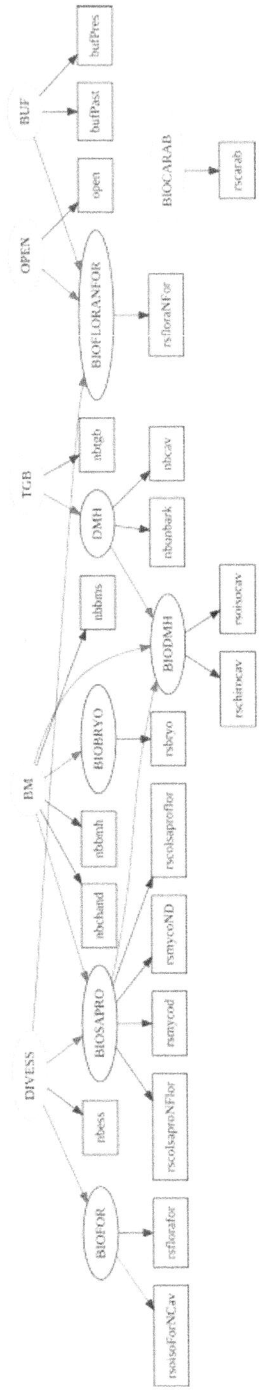

Figure 10.7. *SEM resulting from correction of the model in Figure 10.5, limiting the degradation of fit with respect to the CFA model. For a color version of this figure, see www.iste.co.uk/peyrard/ecology.zip*

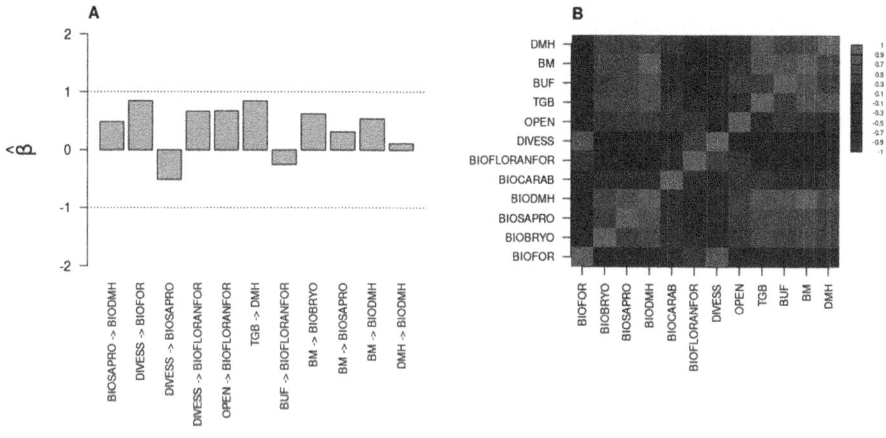

Figure 10.8. *Estimated relationships within the relational model of the SEM from Figure 10.7. Section A: estimators of β type parameters. Section B: estimators of γ type free correlation parameters between exogenous latent variables in the model. For a color version of this figure, see www.iste.co.uk/peyrard/ecology.zip*

Five parameters β are judged to be significant (see Figure 10.9). One notable example is the relation $TGB \rightarrow DMH$, cited as a relationship that was expected to be present on the basis of the theory (see section 10.3.1). Six other parameters are judged to be insignificant. A backward variable selection process could be used in this case to evaluate the extent to which the elimination of some of these relationships improves the parsimony of the model, without significant degradation of fit with respect to the confirmatory analysis model.

We see from Figure 10.9 that the asymptotic significance threshold given by the χ^2 distribution is systematically lower than the threshold obtained by resampling, suggesting that the resampling approach tends to detect fewer significant relationships than the asymptotic approach and that, as a consequence, the asymptotic approach would exhibit an inflated level of type I errors for these tests.

10.3.5. *Findings*

In this chapter, we have presented a step-by-step implementation of the SEM framework using a dataset with few observations and a large number of variables. We simplified a larger dataset, taken from a study by Larrieu *et al.* (2019), restricting the spatial extent and the range of managed forest ecosystems. The biodiversity variables used here were also different from those in the original study, and were normalized, centered and reduced. Thus, the results obtained here are not directly comparable, in quantitative terms, with those of the original study. Nevertheless, there are a number of qualitative similarities. For example, the presence of deadwood has a positive effect (direct or indirect) on the biodiversity of several forest taxons (see Figures 10.7 and 10.8), although a more thorough backwards selection approach could be applied in our case (see Figure 10.9).

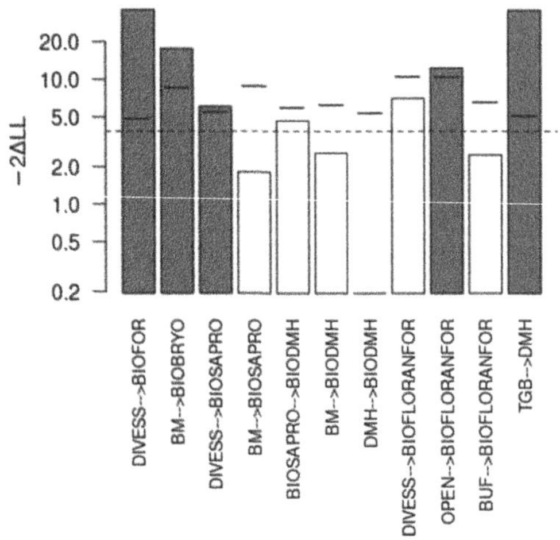

Figure 10.9. *Test of causal relations in the relational model shown in Figure 10.7. Each vertical bar corresponds to the statistic of the likelihood ratio observed when the target causal relationship is neutralized. Red bars correspond to a statistic which is significant at a level of* 0.05*; the thresholds (continuous blue horizontal lines) are determined by resampling (*$B = 100$*). The dotted blue horizontal line corresponds to the asymptotic threshold associated with a* χ_1^2 *distribution. For a color version of this figure, see www.iste.co.uk/peyrard/ecology.zip*

The SEM obtained in this chapter gives a parsimonious description of relations within a dataset. The 21 observed variables were condensed into 12 latent variables, representing the components of managed deciduous forest socio-ecosystems at low altitude. Our relational model shows a simplified structure of the relationships between components, with $6 \times 5/2 = 15$ γ parameters and 11 β parameters, for a total of 26 links in the final model, down from a total of $12 \times 11/2 = 66$ possible links. It also has the capacity to report effects with larger magnitude than those suggested by pairwise relationships among observed variables, through a control of measurement errors, which can be estimated using redundancy between variables within the same group. For example, considering the relationship between the latent variables representing the availability of deadwood in a stand (BM) and bryophyte biodiversity ($BIOBRYO$), the correlation coefficients between each of the observed variables associated with BM and the observed variable for $BIOBRYO$ are between 0.46 and 0.57. The single link expressing this relationship in the final relational model is actually stronger, with an estimated effect of $\hat{\beta} = 0.62$ (see Figure 10.8).

Finally, this approach improves our understanding of managed deciduous forest socio-ecosystems at low altitude, highlighting new relationships, including potential interactions between biodiversity components ($BIOSAPRO \rightarrow BIODMH$). Further discussion of the distinctions between habitat effects and interactions between species in datasets can be found in Chapters 7 and 8, which present other methods based on latent variables that are designed for use in this context.

10.4. Discussion

SEMs provide a statistical framework for compact representations of theories concerning the relations between components in a socio-ecosystem. These components are represented by latent variables. In this chapter, we have shown how SEMs (i) maintain statistical power in cases with large numbers of observed variables, combining variables into groups and linking them to subjacent components through a measurement model; (ii) contribute to improved understanding of socio-ecosystems via progressive adjustments to the SEM structure throughout the analytical process. The approach presented here differs in a number of ways from the wealth of existing literature on SEMs (Bollen 1989; Grace 2006); these differences will be presented and justified below.

10.4.1. *A confirmatory approach*

The approach presented here is, to a great extent, confirmatory, particularly in the case of the example described in section 10.3: we aim to evaluate whether or not an *a priori* model is corroborated by observations. Nevertheless, SEMs can also be used in an exploratory manner, for example through *exploratory factorial analysis* (see Bollen 1989, pp. 226–232), which is used to identify groups of variables giving the best possible compromise between fidelity to the correlation matrix of observed variables and parsimony. For more details on component-based regression methods, see Chapter 9, which focuses on a similar objective.

The example presented here does not feature a strictly confirmatory approach. The residual correlation matrices (Figure 10.6) of a badly fitted SEM were examined in order to identify means of improving the fit, while maintaining a critical approach to added relations on the basis of theoretical knowledge. In practice, therefore, we used a mixed *model generating* approach, similar to that described by Jöreskog and Sörbom (1996) (see also Grace 2006, p. 134).

10.4.2. *Gaussian framework*

In this presentation, the observed variables were presumed to follow a Gaussian vector distribution. In such cases, the variables in a dataset may need to be transformed in order to obey the distribution. It is important to note that this transformation is not always possible, and that the transformation of original variables may also result in a loss of significance in the values of the transformed variables. While the Gaussian hypothesis results in a simple framework, which is easy to implement, notably due to the remarkable compactness and low calculation requirements of the associated likelihood, it can be problematic, notably in cases where quantitative predictions are required. It does not, however, have a negative impact on understanding systems. The application of monotone transformations to data in order to obtain Gaussian distributions does not affect the ordination of values, meaning that SEMs correctly describe whether or not variables are covariant, and whether co-variations are positive or negative. The relationships expressed in the relational model clearly highlight dependency and independence connections between components in a system.

Various methods have been put forward for adapting estimation and testing procedures in order to take account of data, which deviate from the Gaussian vector hypothesis (Satorra and Bentler 1988). However, in cases where greater flexibility in terms of distributions is desirable, it may be better to adopt a more general hierarchical framework in conjunction with a Bayesian approach, although this comes with a higher computation cost.

10.4.3. *Centered-reduced observed variables*

We chose to work with a centered-reduced data set and centered-reduced observed variables in our SEMs. This implies that the variances of non-reduced variables predicted by the SEMs must be equal to the corresponding empirical variances in the non-reduced data set. Our analysis, therefore, focused on the covariance structure between observed variables conditional on the observed variances. This choice is similar to that used, historically, in the context of path analysis (Wright 1921), and draws on the fact that an estimation of the variances of observed variables is of limited interest, given the lack of clear quantitative significance in the transformed variables. Nevertheless, there is no obligation to use this hypothesis, and the variances of the observed variables can also be estimated using the SEM approach.

10.4.4. *Structural constraints*

The structures of the relational and measurement models presented here are subject to more constraints than are strictly necessary to obtain a valid SEM. Some of our restrictions were designed to reduce the risk of losing identifiability and to make the results and diagnoses easier to interpret: namely, the condition that measured variables must relate to only one subjacent variable, and the elimination of latent variables without at least one associated observed variable. While these limitations can, theoretically, be overcome, their use is highly recommended in practice. In our case study, we also introduced free correlation terms γ between all of the exogenous variables, but no others. Again, this choice is not obligatory, and in theory, free correlation terms may be introduced in isolation between any pair of latent or observed variables. However, our constraint makes it possible to project the matrix Σ_F onto the space of positive definite matrices in a simple, effective manner.

10.4.5. *Use of resampling*

In this chapter, we have highlighted testing techniques based on resampling in order to avoid asymptotic hypotheses, which are not suited to our context of study. This focus has inevitable implications in terms of our discussion of SEMs and their simplicity. Estimating an SEM is a numerical optimization operation, and can take several minutes using a standard computer. Multiple resamplings can thus be costly in terms of computation time, and parallelization may be required.

The move from an asymptotic approach to resampling has a major impact on the results of the different tests described here. For example, the asymptotic fit test applied to the final SEM in our case study gives p-value of 4×10^{-6}, suggesting that the SEM should be rejected on the grounds of insufficiency; however, this same SEM passes the resampling test (see section 10.3.3.2). In individual tests of the relations included in the SEM, more relations tend to be considered as significant using the asymptotic approach than when using the resampling approach (see Figure 10.9). These two results suggest that the asymptotic approach has an overall bias in favor of complex models at the expense of simpler forms; this observation illustrates the statistical implications of a move toward non-asymptotic approaches.

10.5. Acknowledgments

The authors wish to thank M. San Cristobal for initiating this project and recruiting the writing team. Thanks are also due to A. Vialatte and J. Rivers-Moore for their valuable contributions to discussions in the course of the writing process. This chapter draws on work carried out as part of the ANRJCJC BloBiForM (ANR-19-CE32-0002-01) project on the development of block analysis techniques for biodiversity in forest environments.

10.6. References

Benjamini, Y. and Hochberg, Y. (1995). Controlling the false discovery rate: A practical and powerful approach to multiple testing. *Journal of the Royal Statistical Society, Series B*, 57(1), 289–300.

Bollen, K.A. (1989). *Structural Equations with Latent Variables*. John Wiley and Sons, New York.

Bollen, K.A. and Stine, R.A. (1992). Bootstrapping goodness-of-fit measures in structural equation models. *Sociological Methods and Research*, (21), 205–229.

Box, G.E.P. and Cox, D.R. (1964). An analysis of transformations. *Journal of the Royal Statistical Society, Series B*, 26(2), 211–252.

Grace, J.B. (2006). *Structural Equation Modeling and Natural Systems*. Cambridge University Press, Cambridge.

Holm, S. (1979). A simple sequentially rejective multiple test procedure. *Scandinavian Journal of Statistics*, (6), 65–70.

Jöreskog, K. and Sörbom, D. (1996). *LISREL 8: User's reference guide*. Scientific Software International, Chicago.

Larrieu, L. and Gonin, P. (2008). L'indice de biodiversité potentielle (JBP) : une méthode simple et rapide pour évaluer la biodiversité potentielle des peuplements forestiers. *Revue Forestière Francaise*, (60), 727–748.

Larrieu, L., Gosselin, F., Archaux, F., Chevalier, R., Corriol, G., Dauffy-richard, E., Deconchat, M., Gosselin, M., Ladet, S., Savoie, J.-M., Tillon, L., Bouget, C. (2019). Assessing the potential of routine structural and dendrometric variables as potential habitat surrogates from multi-taxon data in European temperate forests. *Ecological Indicators*, (104), 116–126.

Lei, P.-W. and Wu, Q. (2015). Estimation in structural equation modeling. *Handbook of Structural Equation Modeling*. The Guilford Press, New York.

Lejeune, M. (2010), *Statistiques : la théorie et ses applications*, 2nd edition. Springer, Springer-Verlag, Paris.

Muirhead, R. (1982). *Aspects of Multivariate Statistical Theory*. John Wiley and Sons, New York.

Pearl, J. (2000). *Causality: Models, Reasoning and Inference*. Cambridge University Press, Cambridge.

Rao, C. (1973). *Linear Statistical Inference and Its Applications*, 2nd edition. John Wiley and Sons, New York.

Satorra, A. and Bentler, P.M. (1988). Scaling corrections for chi-squared statistics in covariance structure analysis. In *Proceedings of the American Statistical Association*. American Statistical Association, Alexandria, VA, 308–313.

Sokal, R.R. and Rohlf, F.J. (1995). *Biometry: The Principles and Practice of Statistics in Biological Research*, 3rd edition. W. H. Freeman and Company, New York.

Wright, S. (1921). Correlation and causation. *Journal of Agricultural Research*, (20), 557–580.

List of Authors

Julyan ARBEL
University of Grenoble Alpes
Inria
CNRS
Grenoble INP
Laboratoire Jean Kuntzmann (LJK)
France

Julie AUBERT
Paris-Saclay University
AgroParisTech
INRAE
UMR MIA-Paris
France

Pierre BARBILLON
Paris-Saclay University
AgroParisTech
INRAE
UMR MIA-Paris
France

Olivier BONNEFON
Biostatistique and Processus Spatiaux (BioSP)
INRAE
Avignon
France

Xavier BRY
Institut Montpelliérain Alexander Grothendieck
University of Montpellier
CNRS
France

Mathieu BUORO
ECOBIOP
INRAE
Saint-Pée-sur-Nivelle
France

Daria BYSTROVA
University of Grenoble Alpes
Inria
CNRS
Grenoble INP
Laboratoire Jean Kuntzmann (LJK)
France

Jocelyn CHAUVET
INSERM U1219 BPH
University of Bordeaux
and
Centre de recherche de l'ICES
La Roche-sur-Yon
France

Pierre-Olivier CHEPTOU
CEFE UMR 5175
CNRS
University of Montpellier
Université Paul-Valery
France

Julien CHIQUET
Paris-Saclay University
AgroParisTech
INRAE
UMR MIA-Paris
France

Stéphane CORDEAU
Agroécologie
AgroSup Dijon
INRAE
University of Bourgogne
Bourgogne Franche-Comté University
Dijon
France

Guillaume CORNU
CIRAD
UPR Forêts et Sociétés
Montpellier
France

Marie-Josée CROS
University of Toulouse
INRAE UR MIAT
Castanet-Tolosan
France

Sarah CUBAYNES
CEFE
University of Montpellier
CNRS, EPHE-PSL University, IRD
Paul Valéry Montpellier 3 University
France

Sophie DONNET
Paris-Saclay University
AgroParisTech
INRAE
UMR MIA-Paris
France

Marie-Pierre ETIENNE
Institut Agro
Agrocampus Ouest
CNRS, IRMAR – UMR 6625
Rennes
France

Jérémy FROIDEVAUX
UMR 1201 Dynafor
University of Toulouse
INRAE INPT EI PURPAN
Castanet-Tolosan
France

Olivier GIMENEZ
CEFE
University of Montpellier
CNRS, EPHE, IRD
Paul Valéry Montpellier 3 University
France

Pierre GLOAGUEN
Paris-Saclay University
AgroParisTech
INRAE
UMR MIA-Paris
France

Michel GOULARD
UMR 1201 Dynafor
University of Toulouse
INRAE INPT EI PURPAN
Castanet-Tolosan
France

Valentin JOURNÉ
Grenoble Alpes University
INRAE, LESSEM
Saint-Martin-d'Hères
France

Etienne KLEIN
Biostatistique and Processus Spatiaux (BioSP)
INRAE
Avignon
France

Fabien LAROCHE
UMR 1201 Dynafor
University of Toulouse
INRAE INPT EI PURPAN
Castanet-Tolosan
and
INRAE
UR EFNO
Nogent-sur-Vernisson
France

Laurent LARRIEU
UMR 1201 Dynafor
University of Toulouse
INRAE INPT EI PURPAN
Castanet-Tolosan
and
CNPF-CRPF Occitanie
Tarbes
France

Valentin LAURET
CEFE
University of Montpellier
CNRS, EPHE, IRD
Paul Valéry Montpellier 3 University
France

Sebastian LE COZ
University of Toulouse
INRAE, UR MIAT
Castanet-Tolosan
France

Julie LOUVRIER
Department of Ecological Dynamics
Leibniz Institute for Zoo and Wildlife Research
Berlin
Germany

Mahendra MARIADASSOU
Paris-Saclay University
INRAE
MaIAGE
Jouy-en-Josas
France

Vincent MIELE
Laboratoire de Biométrie et Biologie Évolutive
Université Lyon 1
CNRS
UMR5558
Villeurbanne
France

Frédéric MORTIER
CIRAD
UPR Forêts et Sociétés
and
Forêts et Sociétés
University of Montpellier
CIRAD
Montpellier
France

Julien PAPAÏX
Biostatistique et Processus Spatiaux (BioSP)
INRAE
Avignon
France

Nathalie PEYRARD
University of Toulouse
INRAE, UR MIAT
Castanet-Tolosan
France

Giovanni POGGIATO
University of Grenoble Alpes
Inria
CNRS
Grenoble INP
Laboratoire Jean Kuntzmann (LJK)
and
University of Savoie Mont Blanc
Laboratoire d'Ecologie Alpine (LECA)
France

Stéphane ROBIN
Paris-Saclay University
AgroParisTech
INRAE
UMR MIA-Paris
France

Lionel ROQUES
Biostatistique and Processus Spatiaux (BioSP)
INRAE
Avignon
France

Nina SANTOSTASI
Department of Biology and Biotechnologies "Charles Darwin"
University of Rome La Sapienza
Italy

Samuel SOUBEYRAND
Biostatistique and Processus Spatiaux (BioSP)
INRAE
Avignon
France

Wilfried THUILLER
University of Grenoble Alpes
CNRS
University of Savoie Mont Blanc
Laboratoire d'Ecologie Alpine (LECA)
France

Catherine TROTTIER
Paul-Valéry Montpellier 3 University
and
Institut Montpelliérain Alexander Grothendieck
University of Montpellier
CNRS
France

Emily WALKER
Biostatistique and Processus Spatiaux (BioSP)
INRAE
Avignon
France

Index

Printed and bound by CPI Group (UK) Ltd, Croydon, CR0 4YY

13/09/2026

14960994-0002